# Anatomy & Physiology

## FOR

# DUMMIES®

# Anatomy & Physiology
## FOR
## DUMMIES®

by Donna Rae Siegfried

WILEY

Wiley Publishing, Inc.

**Anatomy & Physiology For Dummies®**

Published by
**Wiley Publishing, Inc.**
111 River St.
Hoboken, NJ 07030-5774
www.wiley.com

*Library of Congress Cataloging-in-Publication Data:*

Library of Congress Control Number: 2002102446

ISBN: 978-0-7645-5422-3

Manufactured in the United States of America

15   14   13   12   11   10

 is a trademark of Wiley Publishing, Inc.

# About the Author

**Donna Rae Siegfried,** author of *Biology For Dummies* (Wiley Publishing, Inc.), has been writing and editing medical information for 14 years. She began her college years at Moravian College in Bethlehem, Pennsylvania, with the goal of attending medical school and then discovered her knack for writing about biology and medicine. The field of science writing was in its infancy then, but while working at nearby Lehigh University, she found "science writing" in one of Lehigh's course listings. Her career path became clear at that moment. She added journalism to her biology major, took the science writing and medical ethics courses at Lehigh, and began to write.

Donna's journalism internship was done at Rodale Press in Emmaus, Pennsylvania, where she worked with the editor of gardening books. That led to a full-time job with *Organic Gardening* magazine before her college graduation. However, as much as Donna loved publishing and writing about biology topics, gardening (prior to home ownership) just wasn't her forte. Donna became an information analyst at Rodale, where she read about 500 medical journals every month (more her "thing"), selected and abstracted important articles for Rodale's files, created biweekly newsletters, and wrote special reports (such as how supplying vitamin A to third-world countries can dramatically decrease the incidence of blindness in those countries). She began to write small bits for *Runner's World* magazine and then left Rodale Press for an opportunity at a medical publishing company, the pursuit of a master's degree in science and technical communication, and the creation of a marriage.

Donna attended Drexel University in Philadelphia and worked for Williams & Wilkins, Inc. in Media, Pennsylvania, which is now Lippincott/Williams & Wilkins (Philadelphia). Donna worked on staff at Williams & Wilkins as a development editor, working directly with authors of textbooks to add to, change, or correct their manuscripts, making them fit the format of the National Medical Series of review books. She traveled to medical conferences and conducted focus groups of medical students where she gathered more information on how to improve the NMS books and kept up with changes in the United States Medical Licensing Examination (USMLE).

After 5 years on staff at Williams & Wilkins, Donna, her husband, and their then 18-month-old son moved to the tiny mountain village of Germania (Potter County), Pennsylvania, to try the work-at-home lifestyle. There, Donna launched her freelance career and company, Synergy Publishing Services. She develops and edits books for several different medical publishing companies and has written articles on the drugs Avonex and Copaxone for those with multiple sclerosis, the heart surgery technique called transmyocardial laser revascularization, and some alternative medicine treatments.

She has edited dozens of basic science and clinical medicine books and articles (and thousands of USMLE-type questions!). While doing so, she has been fortunate enough to work with some of the leading MDs and PhDs in the country.

Donna also has held a position as an instructor of anatomy and physiology at the Pennsylvania College of Technology in Wellsboro, Pennsylvania. She discovered that she absolutely loved teaching science as much as writing and editing the information, and was just about to start a master's in education program when she, her husband, three children, and two dogs relocated to the Atlanta suburb of Alpharetta, Georgia. There, Donna is a member of two tennis teams and is the flute player in a woodwind group. She still plans to pursue that master's in education and works to promote science education and wellness and to make science fun for kids so that they may choose an area of science for their careers.

# Dedication

I was about to begin writing this book one gorgeous Tuesday morning in September. My first book, *Biology For Dummies*, was at the printing press, and it was time to get to work on this second book. But my phone rang, and my husband told me to put on the news because the World Trade Center had just been hit by an airplane. As I turned on the TV, I saw the second plane hit the other tower. The strangest feeling came over me at that moment. Looking back, I think I knew life would not be the same for any of us from then on. I was watching the news coverage while on the phone with my sister, Melissa Jurnock, who said, "It looks like there was another explosion." I remember saying, not believing it myself, "No. I don't see the building anymore. I think it collapsed."

Having grown up about an hour's drive from New York City, I know of several people who were in the towers when the plane hit. Some survived but are traumatized. And I have heard of many others known by people in my life who died on that ironically beautiful morning. I also know of firefighters who had responded to the initial call and survived, and others who responded to the need to clear the rubble and look for bodies. My own husband has been a volunteer firefighter for nearly 20 years, so I know too well the feeling of dread the wife of a firefighter gets when her husband and father of her children goes out on a bad call. I can only imagine the dread and fear the wives of those firefighters felt that day. I empathize with the families of those innocent victims, and I feel the pride of the families whose loved ones went to help.

So, I dedicate this work, which began several weeks later than originally planned, to the victims and heroes of the September 11 tragedies. I dedicate this book to two longtime friends: Michael Escott, who, fortunately, was not in New York City that morning, but knows many who were lost; and Michael Deegan, whose brother had to evacuate Manhattan and whose stepfather and uncle are two of the brave firefighters who went above and beyond their duty. This tragedy put me back in touch with those two wonderful people, both of whom have had enormous impacts on my life. I dedicate this book to Jo-Ann Fleishman, with whom I used to share an office and writing responsibilities but who was more than a coworker. She became a dear friend and has been an inspiration to me for many years now. Her brother-in-law was in the World Trade Center when the first plane hit. Thankfully, he survived. Peter Fleishman still has his brother; Haley and Andrew still have their uncle.

Like most Americans, I have shed many tears over the enormity of destruction that day. My heart sinks every time I think of the little children — the same ages as my own — who were on those planes or the ones who lost their mommy, daddy, or other loved ones. I dedicate the months of work on this book to them and to the little girl at my church who collected 1,400 stuffed animals and sent them to the children in New York City who are grieving. I dedicate this book to the renewed spirit of Americans and to any good that can come from this horrible tragedy.

I dedicate my work to my own children — to Abby and Ryan — who didn't understand why Mommy was crying a lot in September, and to Steven, who has shown a great sense of patriotism even though he has been scared that something bad will happen to someone in our family. I dedicate it to my niece, Marissa, who had her arm set in a cast on September 11. I hope she remembers September 11, 2001, as the day her arm was broken instead of the day America's heart was broken.

And I dedicate this work to the hope for true world peace — not just cease-fires and temporary truces, but tolerance and respect for other cultures and opinions, and for compromise when there are differences. For understanding that everyone prays to the same God, even though the way that humans have organized the religions devoted to that same God are different — not wrong, just different. And for understanding that women and people of differing ethnic groups around the world are no less human nor less important than any other group.

I can tell you what parts are inside the human body and how they work, but, I can't begin to fathom what is in the hearts and souls of terrorists who think they are doing the right thing by killing innocent people in the name of God. I can tell you how their brains work, but not their minds. I can only hope, like people in the rest of the civilized world, that they will become less fervent and more rational so that honest discussions can solve these problems instead of war. I can only hope that they soon realize how amazing life can be and how special each human life is. God Bless America — and the entire world.

— *Donna Rae Siegfried*

# Author's Acknowledgments

My first thank you is to my family, who once again sacrificed me as wife and mother to the role of author. Many readers probably don't realize how much time it takes to write and produce a book. The *For Dummies* series books have very tight deadlines, which means that you have to write almost every waking moment from the beginning of the project until the end. My kids missed out on several months of being held by their mom and hearing bedtime stories, mom taking them to the park after school, mom making dinner, and mommy being able to play with them during the day. Their dad picked up the slack, though, as he always does, and he was usually asleep when I stopped working late at night. So he missed out on many nights of hearing "thank you" from me. Now is my chance to say thank you to all of them: my husband, Keith, and my children, Steven, Ryan, and Abby. I also thank my neighbors (and friends) Vicky King, Kathryn Ericson, and Jan Pix for having my kids at their houses so I could work; and I thank the many other neighbors who offered. I also thank my friends A.J. Peiffer, Lynn Reinhart, Beverly Gray, Colleen Marx, Patty Reiss, Kristin Weger, Kelly Totten, and Elise Armstrong; everyone on my Medlock Bridge tennis team; the music director at our church, John O'Neal, who easily reduced my writing stress with some challenging and beautiful flute music for me to play; my sister, Melissa Jurnock; my mom, Gail Bonstein; my father, Ray Male; my aunt, Barbara Barr; my sisters-in-law Christine Siegfried and, believe it or not, Christine Siegfried, all of whom provided me with words of encouragement and understanding when I needed it most. What a great support system!

Professionally, I would like to thank my agent, Sue Mellen, for putting up with me through another one of these projects, and her assistant, Barb Cahoon, for keeping me informed. Many thanks to the great folks at Wiley: Roxane Cerda, for getting the project going and helping me hunt down art sources; Linda Brandon and Marcia Johnson, for handling the editing; John Langdon, for doing the technical review; and Kathryn Born, for preparing some of the figures. Also, I would like to thank several fine editors with whom I worked in the past. All of them have greatly contributed to my writing and editing skills, and I want to let them know how much I enjoyed working with them and wish them well now: Claire Kowalchik, Jo-Ann Fleishman, Jane Velker, Debra Dreger, Jane Edwards, Susan Keller, Melanie Cann, Amy Dinkel, Beth Goldner, Laurie Forsyth, Susan Kelly, Matt Harris, and Jim Harris.

## Publisher's Acknowledgments

We're proud of this book; please send us your comments through our online registration form located at www.dummies.com/register.

Some of the people who helped bring this book to market include the following:

*Acquisitions, Editorial, and Media Development*

**Project Editors:** Linda Brandon, Marcia L. Johnson

**Acquisitions Editor:** Roxane Cerda

**Copy Editor:** Esmeralda St. Clair

**Technical Editor:** John Langdon

**Editorial Managers:** Christine Meloy Beck, Jennifer Ehrlich

**Editorial Supervisor:** Michelle Hacker

**Editorial Assistant:** Nivea C. Strickland

**Cover Illustration:** Kathryn Born, MA

*Composition Services*

**Project Coordinator:** Regina Snyder

**Layout and Graphics:** Kelly Hardesty, LeAndra Johnson, Brian Massey, Jackie Nicholas, Shelly Norris, Laurie Petrone, Brent Savage, Julie Trippetti, Jeremey Unger, Mary Virgin

**Special Art:** Kathryn Born

**Proofreaders:** John Greenough, Andy Hollandbeck, Carl Pierce, Linda Quigley, Dwight Ramsey

**Indexer:** Aptara

*Special Help:*
E. Neil Johnson, Tonya Maddox, Patricia Yuu Pan

*Publishing and Editorial for Consumer Dummies*

**Diane Graves Steele,** Vice President and Publisher, Consumer Dummies

**Joyce Pepple,** Acquisitions Director, Consumer Dummies

**Kristin A. Cocks,** Product Development Director, Consumer Dummies

**Michael Spring,** Vice President and Publisher, Travel

**Brice Gosnell,** Publishing Director, Travel

**Suzanne Jannetta,** Editorial Director, Travel

*Publishing for Technology Dummies*

**Richard Swadley,** Vice President and Executive Group Publisher

**Andy Cummings,** Vice President and Publisher

*Composition Services*

**Gerry Fahey,** Vice President of Production Services

**Debbie Stailey,** Director of Composition Services

# Contents at a Glance

# Cartoons at a Glance

*By Rich Tennant*

page 9

"Okay, you're really catching on to where all the body parts go, but we need to remember that there's quite a bit going on behind the scenes. Now, we're going to get good and gory and study the innards!"

page 133

"NOW THAT I'VE LIGHTENED UP THE ROOM,..."

page 65

"I may not know anything about genetics, but I know that I'm overdue and it's your side of the family that's always late for family functions."

page 275

"Frank have used to teach high school physiology, so if you value your zygomatic arch or your Alveolar margine, you'll start talking."

page 321

**Cartoon Information:**
**Fax:** 978-546-7747
**E-Mail:** richtennant@the5thwave.com
**World Wide Web:** www.the5thwave.com

# Table of Contents

# Introduction

*W*elcome to *Anatomy & Physiology For Dummies,* your personal owner's manual to, well, your person! Everyone should have such a guide. You probably have an owner's manual for your car, your grill, your DVD player, and even your phone. Why not one for your body? After all, your body is the one "machine" that you use constantly throughout your life. You should understand where the parts of the body are, what they do, and how its systems work together.

Of course, the more that you read about anatomy and physiology, the more the information gels and makes sense. Your body goes through recurring cycles and processes every day to keep you functioning. After reviewing these cycles and processes a few times, the way the body works may suddenly seem so simple — yet amazing. Timing isn't everything, however, when it comes to how your body works. The structure and the location of your body parts also play a role.

The human body is a fascinating place — a work of art, even. Get into the body for a while and figure out how yours works. If nothing else, I hope this knowledge makes you appreciate how special you are and why (and how) you should take care of your body.

## About This Book

As complicated as the human body may appear at first glance, you can easily look at it system by system and part by part. Doing so takes the mystery out of certain activities that your body does on a daily basis, such as breathing, eating, excreting waste, pumping blood, and healing cuts. Each system has a distinct purpose, and many systems of the body work together. Most of the chapters of this book focus on a particular body system. You also find chapters that give you ten ways to take care of your body and ten places on the Web to find more information.

# Conventions Used in This Book

The format of most traditional anatomy books is the same: Chapters focus on systems of the body and explain where the important structures are and how they function. This book is totally conventional in that sense. Anatomy books are organized that way because the human body is organized that way. I have no right to change either.

Well, I shake up a few traditions in this book. I don't use many anatomical *pointer* terms — the drab directive language that is typical in older anatomy books. The classics may include more than you ever wanted to know about anatomy, but you won't need to count sheep with language like this: " . . . is situated at the posterior part of the bone . . . gives attachment to the occipito-frontalis muscle, is perforated by numerous foramina, transmits a vein to the lateral sinus and a small artery from the occipital to supply the dura mater."

I use simple language. While teaching an introductory anatomy and physiology class to mostly adult learners, I found that using "directive" language didn't help if students were trying to find out where in the body certain structures are located. It was like using an old map to tell someone how to find buried treasure — students tuned out before the description came to an end! Instead, I use regular descriptions and language: *in front of, underneath, to the right of,* and so on. Using these kinds of words should make it easier for you to figure out where structures are located. You have to promise me that you'll do two things, however. In order for the simplified language to work, you have to (1) be thinking in terms of anatomical position (see Chapter 1); and (2) you have to look at the figures. If you do those two simple things, you'll enjoy this book more.

Also, I haven't loaded up the text with Latin words. Anatomy and physiology are full of Latin and Greek terms or terms derived from Latin and Greek words. So where appropriate, I explain the Latin or Greek root so you can better understand the term(s). This change, too, should make your anatomy reading experience less cumbersome and more enjoyable.

# What You're Not to Read

Don't feel obligated to read every word that I've written (unless you're my mother). The sidebars, for example, are meant to supplement the information in the text. The sidebars are not required reading and aren't essential to your understanding the rest of the material in the chapter.

# Foolish Assumptions

I am assuming that you're one of three people:

✔ A high school student preparing for an advanced placement test or college entrance examination

✔ A college student trying to make sense of or review the tons of material you learned throughout an anatomy and physiology course

✔ An adult learner who is taking an anatomy and physiology course, considering taking an anatomy and physiology course, or trying to get through a course

Perhaps you're none of those three; you could just be the proud owner of a human body who wants to know how that fantastic machine works. Perhaps you have seen many different types of human bodies and you wonder how differences happen. Or maybe you wonder exactly how the foods you eat, how much you drink, and if you exercise really do affect your health. But, it is possible, too, that the circle of life seems so final, yet infinite at the same time, and you want to know more about how it continues.

Whatever your reason for picking up this book, I have done my best to explain the topics of anatomy and physiology simply and effectively. I hope it works for you!

# How This Book Is Organized

The yellow and black book that you are now holding has a simple arrangement. Before you get into the "meat" of anatomic structures, the first part of this book includes chapters covering the background of anatomy and physiology: anatomic position, increasingly magnified aspects of the body, what the body must do regularly to survive, and the divisions of the body. Then the anatomy part — arranged just like the human body from the inside out — explores bones, muscles, and the skin that covers them. In the part on physiology, you discover each of the body's organ systems, parts included. In both the anatomy and physiology parts, you can find out *how* structures function and how they relate to other parts of the body. Continuing the circle of life commands its own part of the book, with chapters on reproduction, birth, and development. The final part of the book provides you with some helpful information to make the experiences of having a human body or learning more about the human body more interesting.

# Part 1: Positioning Yourself to Study Anatomy

Before I jump into explaining where in the body certain structures appear, you need to have a reference point. You need to know that you're looking at the human body as if you were looking in a mirror with your palms facing forward. Otherwise, left and right are mixed up. And I wouldn't want you thinking your liver was on your left side!

Also in this part of the book, you take a look at the body as an entire human *organism*. In Chapter 1, you find out that a human is an organism just as a worm is an organism. Organ systems run organisms — any organism — and those organ systems consist of organs, which consist of tissues, which consist of cells, which consist of molecules, which consist of atoms. All cellular organisms are pretty much the same; I go through the body to show you how organisms "differentiate" — become different from one another.

Chapter 2 provides an overview for the physiology chapters that come later in the book. I describe the basic cellular processes that occur in each organism — including humans. These processes include metabolizing, homeostasis (maintaining balance), growing, transferring energy, moving, and reproducing. In this chapter, you also find information on the basics of genetics, such as how chromosomes transfer genetic material during cell division.

In Chapter 3, you develop your anatomy vantage point. You explore the basics of the body's planes, cavities, regions, and membranes, so you'll be ready to head to Part II or Part III to find out all about specific structures and the functions.

# Part II: Anatomy from Head to Toe

The chapters in Part II give you information on the body's skeleton, the muscles that enable the skeleton to move, and the skin that protects the muscles and organs. Also, starting in this part of the book, you'll find that each chapter has a pathophysiology section. *Pathophysiology* is a branch of physiology that deals with how disease develops through changes in the structure or function of a body part or system. I include many of the most common diseases and illnesses that plague the body system under discussion in each chapter. Physiology explains how structures work together and how the processes of the body function; pathophysiology explains what happens when something goes wrong.

# Part III: Focusing on Physiology

This part of the book delves into the body's systems, which is the physiology side of an anatomy and physiology course. Anatomy and physiology courses start off with having the students understand bones and muscles, but the bulk of the course is studying how the systems of the body work together. This section is long, only because the body has many different systems. A major theme of this book is to demonstrate how the systems of the body work together all the time, so it makes sense for the physiology section to remain whole, keeping all the systems of the body in this part. The exception is the reproductive system, which appears in a part devoted to creating a new organism.

# Part IV: Creating New Bodies

Parts II and III focus on the structures and processes that keep one organism (man or woman) living, breathing, and functioning. This part of the book deals with the continuation of the life cycle; that is, how a new organism (baby) is created. The processes of reproduction, birth, and development appear in this part. Development refers to the changes that an organism goes through, and in Chapter 15, the development of a human from before birth until death. Human developmental stages include zygote, embryo, fetus, infant, child, teen, and adult. You find out about the changes that occur as an adult ages on the way to death. You may think all this sounds morbid, but it's so natural.

My sincere hope is that by reading this book, you'll come to realize that you are a part of nature and that you'll appreciate your body now and have plenty of time to enjoy what it can do before your loop through the circle of life is done.

# Part V: The Part of Tens

This fun part gives you two useful chapters. One provides you with ten great ways to keep your body healthy, which I hope you'll want to do after you understand how the body works. The other provides you with ten cool Web sites where you can find more information about anatomy and physiology or quiz yourself on what you did figure out. Have fun!

# Icons Used in This Book

The little round pictures that you see in the margins throughout this book are icons that alert you to neat stuff to know or remember.

The bull's-eye symbol lets you know what you can do to improve your understanding of an anatomic structure, or what you can do to improve your health in that area of the body.

This icon shows where I decipher and explain scientific or technical language in normal words. The symbol also flags extra information that takes your understanding of anatomy or physiology to a higher level, but you don't need it to understand the material in the chapter.

The information next to this icon provides you with interesting tidbits about the body. These icons give you some facts for impressing (or grossing out) people at parties.

This little icon serves to jog your memory. Sometimes the information spotlighted just points out information that I think you should permanently store in your Anatomy and physiology file. Other times, the info here makes a connection between what you're reading and related information elsewhere in the book. If you want a quick review of anatomy and physiology, scan through the book by reading the Remember icons. No need for a chunky yellow highlighter!

# Where to Go From Here

In most *For Dummies* books, you can jump into any part of the book and read snippets here and there as you please. You can do that with this book, too, if you wish. I recommend, however, that you read this book in order. The body is built upon increasingly more complex structures — cells form tissues, which form organs, and then systems and then whole organisms. Similarly, anatomy and physiology information builds upon previous information. Understanding how the immune system works is hard if you don't understand how the circulatory system works. Figuring out how your body exchanges nutrients and oxygen for wastes and carbon dioxide is also diffi-cult if you don't understand why the body needs to do that. Of course, if you already have a basic understanding of metabolism and how the body's struc-tures are arranged, feel free to try reading the book in random order. Another suggestion is to read this book more than once; perhaps read it in order the first time, and then go back and read it in random order thereafter.

The more you read about anatomy, the less complex physiology becomes. After a while, the body doesn't seem complex at all, but instead, becomes an artistically, elegantly organized group of common-sense systems.

# Part I
# Positioning Yourself
# to Study Anatomy

The 5th Wave          By Rich Tennant

# In this part . . .

Before you delve into discovering where body parts are and what they do, build a solid foundation in knowing how the body works. Chapter 1 shows you the inside of your body from smallest (atoms) to largest (organ systems). Chapter 2 goes through a "To Do" list — functions that your body must perform regularly for its own survival and perpetuation of the species: metabolizing foods and oxygen; maintaining balance among systems and with the outside environment; developing and replacing cells as you age; moving; and reproducing. You also find the basics of genetics in the section on replacing cells. Then Chapter 3 shows you how to look at and think about the human body as you begin to find out more about structure and function.

# Chapter 1

# Parts of the Whole

· · · · · · · · · · · · · · · · · · · · · · · · · · · · · · · · · · · · · · · · ·

· · · · · · · · · · · · · · · · · · · · · · · · · · · · · · · · · · · · · · · · ·

"*P*arts is parts," the old chicken commercial goes. What parts of a dead chicken are pressed into a patty may not matter to some people (it matters to me though!), but when you're talking about living, breathing animals, parts are not just parts. Each part of the body has an important and specific function, without which the body doesn't function optimally. Many of the parts work together to keep you up and running.

The study of the structure and location of those parts is *anatomy;* the study of the function of those parts is *physiology.* For example, while studying the anatomy of the heart, you look at the heart's valves, chambers, and blood vessels. Visualizing the structures in the heart makes the physiology of the heart easier to grasp; that is, the details about how the heart passes blood through those valves, chambers, and blood vessels. This chapter gives you an overview of anatomy and physiology and tells you why they're often paired. And this chapter provides information on the proper perspective for looking at the body while you're reading about anatomy. You also get a look at what makes up the body layer by layer.

## Dissecting Anatomy, Physiology, and Pathophysiology

In a nutshell, anatomy is the study of the body's parts. Of course, that's like saying art is simply the use of paint or driving is simply a matter of putting a car in gear.

Physiology goes hand in hand with anatomy. Did you ever hear the adage, "Form follows function?" Well, that pretty much sums up why anatomy and physiology are inseparable. Physiology focuses on the function of body parts from large to microscopic. Anatomy focuses on the form of the organism; that is, all the parts that make up that organism. "Form follows function" implies that parts of the body look the way they do because of the tasks that they need to perform. Generally, that holds true. Various parts have evolved to perform particular tasks for which they're well suited.

This section gives you a quick rundown of the basics of anatomy and physiology so that you can see how these two sciences together help you understand the human body. I also discuss *pathophysiology,* which looks at the chain of events that result from a disease or illness. In doing so, pathophysiologists can suggest ways to improve a patient's health.

## Ann Atomy and her relatives

If you know that the hand is connected to the arm bone, the arm bone is connected to the shoulder bone, the shoulder bone is connected to the collar bone, and the collar bone is connected to the rib bones, then you know anatomy — right? Well only to a degree. Anatomy is a broad subject, and some folks spend their lives trying to understand just one or two little parts of it. Therefore, the discipline of anatomy has several subsets — relatives of "Ann Atomy," if you will. Take a look at the following examples:

- ✔ **Developmental anatomy** focuses on how an individual forms from a fertilized egg all the way through adulthood. Developmental anatomists look at how certain body parts or *systems* (groups of parts that work together) change throughout the life span (see Chapter 15).

- ✔ **Gross anatomy** is the study of large parts of the body that can be seen with the naked eye. Contrary to its title, gross anatomy isn't all blood and gore. Some blood, maybe, but that's to be expected in a body, don't you think? If you took French, you know that *grosse* means large. Therefore, gross anatomy deals with the large parts that can be seen with the naked eye. Gross anatomists aren't mad scientists; they're patient scientists who study each and every detail of the organs, muscles, bones, nerves, and vessels.

- ✔ **Histologic anatomy** is the study of different tissue types and the cells that comprise them. Histologic anatomists use a variety of microscopes to study the cells and tissues that make up the parts of the body. (See Chapter 3 for more on histology.)

The stem *histo-* refers to tissues, as in the branch of science that's dedicated to the study of cells. It comes from the Greek word *histos,* which means web or loom. So think of your tissues as being sheets of cells woven together, and you're on the right track.

# The benefits of a bioengineered fit

If you or anyone you know has a prosthetic arm or leg, you can thank an anatomist for making movement possible in a damaged or diseased limb. Without anatomists, the field of bioengineering — using the principles of engineering for biological or medical purposes — wouldn't exist. Scientists first needed to fully understand every structure of real human bodies inside and out before they could come up with ways of creating replacement parts. These days, arms or legs aren't all that's being fitted for prostheses. Hips, hearts, heart valves, and increasingly smaller parts can be replaced as well. Eyeglasses and contact lenses, which are commonplace, also couldn't have been invented without anatomists figuring out how the eye works. As anatomists discover more, they can pass their information on to other scientists, such as bioengineers, to continue to create more ways to improve the quality of people's lives.

All these branches of anatomy don't focus solely on humans, although some anatomists specialize in human anatomy. Developmental anatomists, gross anatomists, and histologic anatomists can study the bodies and parts of all animals, and their studies are important. The work of anatomists contributes to medical advances, such as improved surgical techniques or the development of bioengineered prostheses. Throughout this book, you encounter some information from each major subset of anatomy.

## The function of physiology

Although each body part seems to function or move on its own, each part usually prompts another body part — this is the study of physiology. For example, moving a hand away from a hot stove takes not only the movement of the hand and arm but also requires the function of the brain and nerves; running requires not only the muscles in your legs but also the bellows motion of your lungs to breathe. If you look at increasingly smaller parts of the body — from organs to tissues, from tissues to cells, and from cells to molecules — you begin to see how more and more parts of the body work together.

Just as several body parts and systems work together to achieve a single result (such as moving, digesting, and reproducing), individual structures or systems of the body also can have more than one function. For example, your blood vessels serve as a network of highwaylike transportation lanes, and the blood and its cells function as 18-wheelers delivering and picking up materials. Blood cells are not only part of the circulatory system; they're integral parts of the respiratory system, digestive system, and immune systems, where they have several physiologic functions described in Table 1-1:

| Table 1-1 | Blood Cell Functions |
|---|---|
| *Process* | *What Blood Cells Do to Help* |
| Respiration | Transport oxygen from the lungs to every other cell in the body |
| Digestion | Carry nutrients derived from eaten foods to all cells in the body |
| Excretion and urination | Carry wastes disposed of by the cells that filter the blood to remove wastes |
| Immune | Transport the cells that fight off organisms that invade the body |

The blood and blood cells work together with several organ systems throughout the body to help keep you functioning properly. Blood is one anatomic entity but it contains several different types of cells, and they have many physiologic functions.

The next section of this chapter shows you the body's basic building blocks that make up the human anatomy and its physiologic functions. Part II of the book goes over the anatomic structures in detail, and Part III focuses on physiology, showing just how the body's systems work together.

# Building the Body: From Atoms to Organs

Your body as a whole is one organism. However, many, many parts make up that whole. As you consider the various levels of the body (see Figure 1-1), you understand that a large number of parts are within parts. It's akin to looking at a pine tree. At first, you notice the entire tree — a whole organism. But, as you look closer, you notice the branches. Looking at the twigs on the branches, you notice each needle on the twigs. Thousands, if not millions, of needles exist on that one single pine tree. The same analogy holds for the human body or the body of any animal. First, you notice the entire body. But the entire body is made up of parts and organs, and each of those organs is made up of a variety of tissues. And if, like a pathologist does, you examine a magnified sample of one of the human body's tissues under a microscope, millions of cells become visible. Yet you can turn up the magnification for an even closer look: Cells contain molecules that are made up of even smaller components called *atoms*.

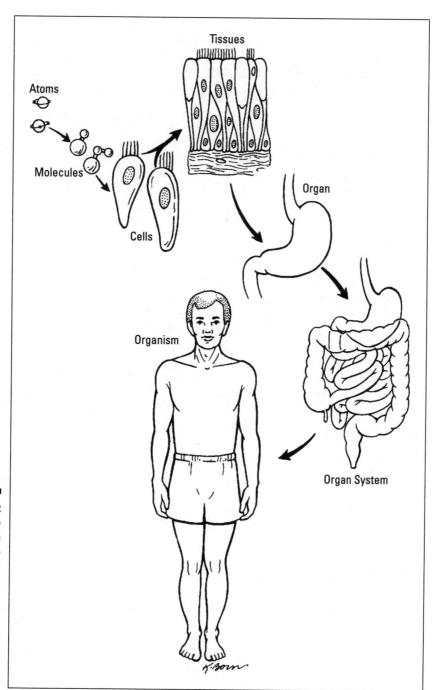

**Figure 1-1:**
Levels of the body from smallest to largest: atoms, molecules, cells, tissues, organs, and organ systems.

## Care for a little chemistry with your anatomy?

Bet you weren't counting on getting a little chemistry with your anatomy book. Well, I'm throwing in some chemistry as a bonus. But if you're really a bargain hunter, you may know that this isn't such a special deal. You see, chemistry is central to many scientific disciplines: obviously chemical sciences but also physical science and life science. As my college chemistry professor said repeatedly, "Chemistry is the central science. That's why the chemistry labs are on the second floor in between the physics floor and the biology floor."

Living things are made up of chemicals; animals and plants are beautiful, wondrous containers for millions of chemical reactions. So I must explain a little bit about what chemicals are and how they react inside your body. Ready?

I bet that what you think of as *chemicals* are really the elements found in the *Periodic Table* (that huge chart that lists the crucial data for each known substance found in, on, and around

Earth). Each substance listed in the Periodic Table is an element found somewhere on our planet — in the air, water, soil, or buried deep underground. I know it may sound like I'm leading to a discussion on geology, but believe me, this has everything to do with anatomy.

Billions of years ago, the planet was covered with many, many active volcanoes, which formed land masses upon the cooling of lava. The gases contained in the volcanic eruptions became inorganic molecules in the land. During this time, the hydrogen released from the volcanic eruptions combined with oxygen in the atmosphere to produce water. Eventually, cells formed from the raw materials of water, earth, and energy. And, over billions of years, living organisms evolved from the cells. (See *Biology For Dummies* by Donna Rae Siegfried [Hungry Minds, Inc.].) But the raw materials that started life are still present in every living thing on earth — animals as well as plants. Those raw materials are the *elements*.

Atoms, molecules, cells, tissues, organs, and organ systems are the body's building blocks. So, in a sense, they're the building blocks for the rest of this book. Getting to know these most basic parts to see how their functions affect the rest of the parts of the body is a good idea.

## Combining atoms to make molecules

An *atom* is the smallest possible piece of an element that retains all the properties of that element. For example, a hydrogen atom reacts the same as a barrel full of hydrogen. Each atom is a building block. If you put two atoms of hydrogen (H) together, you get a molecule ($H_2$). If you add that molecule of hydrogen to an atom of oxygen (O), you create a *molecule* of water. A *molecule* is a conglomerate of atoms. Got that?

Your body contains many different types of molecules that form the working parts such as cells (see the next section) as well as substances produced by some of those parts such as hormones (see Chapter 8). The rest of this book is devoted to showing you all your parts and explaining how they work inside and out.

## Singling out cells: The stuff of life

Your cells perform many important functions without which you wouldn't be able to go about your business. While you breathe, your cells exchange the *bad* air for the *good* air. While you eat, cells produce the *enzymes* (proteins that speed up a chemical reaction) that digest the food and convert the nutrients into a useable form of energy. In short, your cells are like tiny motors that keep you running.

Each cell in your body performs the same tasks that your body as a whole performs:

- ✔ Converting energy
- ✔ Digesting food
- ✔ Excreting waste
- ✔ Reproducing
- ✔ Taking in oxygen

No smaller component than the cell performs all those important functions. That's why the cell is the "fundamental unit of life." Every living thing has cells, and those cells perform basically the same functions whether they're in a human, a horse, or a hyacinth. Turn to Chapter 3 for an in-depth look at cells.

## Teaming up with tissues

The body contains several different types of cells, such as blood cells, nerve cells, and muscle cells. When cells of the same type "hang together" so to speak, and perform the same function, a *tissue* is formed. When you think of *tissue,* you probably think of skin, if not the box of facial tissues that you grab whenever you sneeze. In fact, the body contains four classes of tissues.

- ✔ **Connective tissue,** which is found in blood and bones, serves to support body parts and bind them together.

- ✔ **Epithelial tissue (epithelium)** is the type of tissue that lines organs and covers the body.

- ✔ **Muscle tissue** — surprise! — is found in the muscles, which allow your body parts to move via the acts of contraction and relaxation.

- ✔ **Nerve tissue** transmits impulses and forms nerves.

For more info about the types of tissues and what they do, head to Chapter 3.

# Organs (not the keyboard kind)

Atoms make up molecules; molecules make up cells; cells make up tissues; and two or more kinds of tissues working together make an organ. An *organ* is a part of the body that performs a specialized physiologic function. For example, the stomach contains epithelial tissue, muscle tissue, nerve tissue, and connective tissue, and the stomach has the specific physiologic function of breaking down food. (How the stomach works as part of the digestive system is covered in Chapter 11; other organs are covered throughout Part II and Part III of this book.)

# Organizing your organ systems

An *organ system* is a group of specialized organs working together to achieve a major physiological need. For example, the mouth, esophagus, stomach, small intestine, and large intestine are all organs of the digestive system. The digestive system is the organ system responsible for breaking down foods into nutrients that can be transported through the bloodstream. (See Chapter 11 for more on the digestive system.)

Chapters throughout Part III cover the major organ systems and describe their structures and physiologic functions, as well as what can go wrong in those systems and organs.

# Labeling Your Parts

Science is riddled with Latin terms, and because anatomy is a science, it's no different. Every part of the body has a Latin name. But rather than overwhelm you with Latin terms, Table 1-2 gives you a handy list of some of the most common Latin roots used in anatomy. When you see these roots, you'll have an easier time of figuring out what the term means.

| Table 1-2 | Latin Roots for Common Anatomical Terms | |
|---|---|---|
| *Latin Root* | *Meaning* | *Example* |
| Aden- | Gland | Adenopathy (disease in a gland) |
| Angi- | Vessel | Angioplasty (technique of opening the tissue lining a blood vessel) |
| Arthr- | Joint | Arthritis (inflammation of a joint) |
| Bronch- | Windpipe | Bronchitis (inflammation of tube that carries air from windpipe to lung) |
| Carcin- | Cancer | Carcinogen (substance that causes cancer) |
| Cardi- | Heart | Cardiac arrest (stoppage of heart-beat) |
| Carp- | Wrist | Carpal tunnel syndrome (painful condition in which nerve gets trapped in carpal bones of wrist) |
| Chol- | Bile, gall | Cholesterol (made in liver along with bile) |
| Derm- | Skin | Dermatitis (inflammation of the skin) |
| Erythro- | Red | Erythrocyte (red blood cell) |
| Gastr- | Stomach | Gastric juice (acids and enzymes that digest food chemically) |
| Hemat- | Blood | Hematocrit (count of cells in blood) |
| Histo- | Tissue | Histocompatability (the degree to which a donor's tissues match a recipient's tissues for a graft) |
| Path- | Disease | Pathogen, pathology (see section below) |
| Sept- | Contamination | Septic shock (drop in blood pressure due to contamination of blood) |

# Here's Looking at You, Kid!

I want to make sure that you know where I'm coming from when I use certain terms. If you aren't looking at the body from the correct perspective, you'll have your right and left confused. This section shows you the anatomic position, planes, regions, and cavities, and the main membranes that line the body and divide it into major sections.

## Getting in position

Stop reading for a minute. Stand up straight. Look forward. Let your arms hang down at your sides with your palms facing forward. You are now in *anatomic position* (see Figure 1-2). Whenever you see an anatomical drawing, the body is in that position. Using this position as the standard removes confusion. If one anatomist looks at the body from the back and refers to the right side, and one anatomist looks at the body from the front and talks about the right side, confusion exists. The anatomic position puts everybody on the same page.

The following list of common anatomic descriptive terms that appear throughout this and every other anatomy book may come in handy:

- ✔ **Anterior:** Front or toward the front of the body
- ✔ **Posterior:** Back or toward the back of the body
- ✔ **Dorsal:** Back or toward the back of the body
- ✔ **Ventral:** Front or toward the front of the body
- ✔ **Caudal:** Near or toward the tail
- ✔ **Prone:** Lying on the stomach, face down
- ✔ **Supine:** Lying on the back, face up
- ✔ **Lateral:** On the side or toward the side of the body
- ✔ **Medial or median:** In the middle or toward the middle of the body
- ✔ **Proximal:** Nearer to the point of attachment or the trunk of the body.
- ✔ **Distal:** Farther from the point of attachment or the trunk of the body (think "distance")
- ✔ **Superficial:** Near the surface of the body

- ✔ **Deep:** Farther from the surface of the body
- ✔ **Superior:** Situated above or higher than another part
- ✔ **Inferior:** Situated below or lower than another part
- ✔ **Central:** Near the center (median) of the body or middle of an organ
- ✔ **Peripheral:** Away from the center (midline) of the body or organ

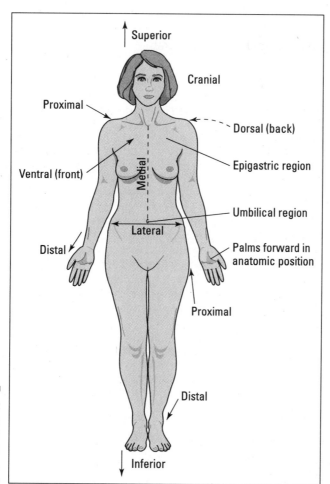

**Figure 1-2:**
The
standard
anatomic
position.

From LifeART®, Super Anatomy 1, © 2002, Lippincott Williams & Wilkins

## Dividing the anatomy

If you ever had a class in geometry, you found out that planes are flat surfaces and a straight line could run between two points on that flat surface. Geometric planes can be positioned at any angle. In anatomy, usually three planes separate the body into sections. Figure 1-3 shows you what each plane looks like. The reason for separating the body into sections — also referred to as *cuts* — is so that you know which *half* of the body is being discussed. The anatomic planes are

- ✔ **Frontal plane:** Divides the body into a front (anterior) portion and a rear (posterior) portion.

- ✔ **Sagittal plane:** This vertical plane divides the body lengthwise into right and left sections. If the vertical plane runs exactly down the middle of the body, it is referred to as the *midsagittal plane.* Otherwise, a sagittal plane can run vertically down through the body at any point, creating a *longitudinal section.*

- ✔ **Transverse plane:** Divides the body horizontally, into top (superior) and bottom (inferior) portions. Dividing horizontally does not necessarily yield two equal divisions; that is, a transverse plane doesn't always go through the waist area to separate the body into top and bottom. Transverse planes can go anywhere to create *cross sections.* When looking at a cross section of a body part, imagine that the body is sectioned horizontally. Or think of a music box that has a top that opens on a hinge. The transverse plane is where the music box top separates from the bottom of the box. Imagine that you open the box by lifting the lid, and you look at the material lining the lid. That's the vantage point that you have when looking at a cross section.

Anatomic planes can "pass through" the body at any angle — the planes are arbitrary for the convenience of anatomists. Don't expect the structures of the body, and especially the joints, to line up or move along the standard planes and axes.

## Mapping out your regions

Three types of planes divide, but regions also separate the body. Just like on a map, a region refers to a certain area. The body is divided into two major portions: axial and appendicular. The axial body runs right down the center (axis) and consists of everything except the limbs, which leaves the head, neck, thorax (chest and back), abdomen, and pelvis. The appendicular body consists of appendages, otherwise known as arms and legs. Each part of the axial and appendicular portions of your body contains regions that are listed in Table 1-3.

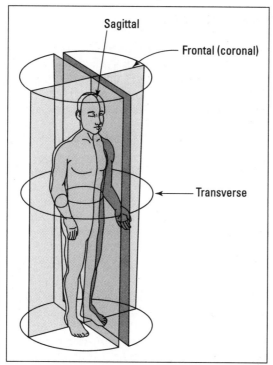

**Figure 1-3:**
Planes of
the body:
Frontal,
transverse,
and sagittal.

*From LifeART®, Super Anatomy 2, © 2002, Lippincott Williams & Wilkins*

| Table 1-3 | Regions of the Body |
|---|---|
| *Axial* | *Appendicular* |
| **Head and neck** | **Arms** |
| cephalic (head) | brachial (upper arm) |
| cervical (neck) | carpal (wrist) |
| cranial (skull) | cubital (elbow) |
| frontal (forehead) | forearm (lower arm) |
| occipital (back of head) | palmar (palm) |
| ophthalmic (orbital, eyes) | |
| oral (mouth) | |
| nasal (nose) | |

*(continued)*

**Table 1-3** *(continued)*

| Axial | Appendicular |
|---|---|
| **Thorax** | **Legs** |
| axillary (armpit) | femoral (thigh) |
| costal (ribs) | lower leg (below the knee) |
| mammary (breast) | pedal (foot) |
| pectoral (chest) | popliteal (back of knee) |
| vertebral (backbone) | |
| **Abdomen** | |
| celiac (abdomen) | |
| gluteal (buttocks) | |
| groin (area of abdomen near thigh) | |
| inguinal (groin) | |
| lumbar (lower back) | |
| pelvic (lower part of abdomen) | |
| perineal (area between anus and external genitalia) | |
| sacral (end of vertebral column) | |

## *Cavities your dentist already knows about*

If you remove all the internal organs, the body is empty except for the bones and tissues forming the space where the organs were. Just as a dental cavity is a hole in a tooth, the body's cavities are "holes" where organs are held (see Figure 1-4). The two main cavities are the *dorsal cavity* and the *ventral cavity.*

The dorsal cavity consists of two cavities that contain the central nervous system. The first is the *cranial cavity,* the space within the skull that holds your brain. The second is the *spinal cavity,* the space within the vertebrae where the spinal cord runs through your body.

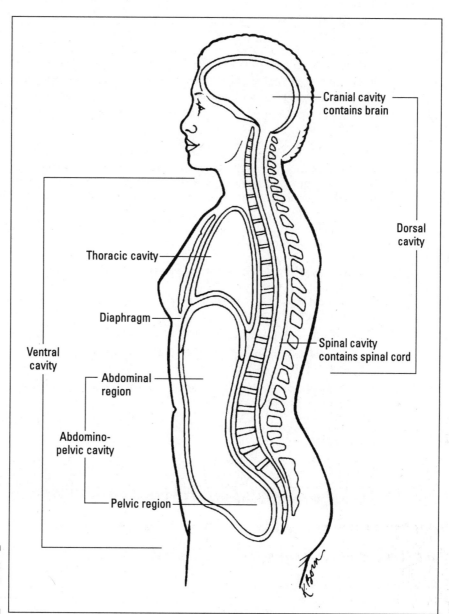

Cranial cavity
contains brain

Dorsal
cavity

Thoracic cavity

Diaphragm

Spinal cavity
contains spinal cord

Ventral
cavity

Abdominal
region

Abdomino-
pelvic cavity

Pelvic region

**Figure 1-4:**
The body's
cavities.

The ventral cavity is much larger and, other than your brain and spinal cord, contains the rest of your organs. This cavity is divided by the diaphragm into smaller cavities: the *thoracic cavity,* which contains the heart and lungs, and the *abdomino-pelvic cavity,* which contains the organs of the abdomen and the organs of the pelvis. The abdominal organs are the stomach, liver, gallbladder, spleen, and most of the intestines. The pelvic cavity contains the reproductive organs, the bladder, the rectum, and the lower portion of the intestines.

Additionally, the abdomen is divided into quadrants and regions. The midsagittal plane and a transverse plane intersecting at an imaginary axis passing through the body at the navel (belly button) divide the abdomen into quadrants (four sections). Putting an imaginary cross on the abdomen creates the right upper quadrant, left upper quadrant, right lower quadrant, and left lower quadrant. Physicians take note of these areas when a patient describes symptoms of abdominal pain.

The regions of the abdominopelvic cavity include the following:

- **Epigastric:** Above the stomach and in the central part of the abdomen just above the navel

- **Hypochondriac:** Doesn't moan about every little ache and illness but lies to the right and left of the epigastric region and just below the cartilage of the rib cage (*chondral* means cartilage, *hypo-* means below)

- **Hypogastric:** Below the stomach and in the central part of the abdomen just below the navel

- **Iliac:** Lies to the right and left of the hypogastric regions near the hip bones

- **Umbilical:** The area around the navel (the umbilicus)

- **Lumbar:** Forms the region of the lower back to the right and left of the umbilical region

# Nothing's Perfect: When Things Go Wrong

Okay. You know that physiology is the study of the functions and processes that occur in the body. But, like life itself, nothing is perfect or guaranteed. Disease happens. The body ages and processes decline in their effectiveness. Health deteriorates when the body's systems experience problems. In Greek, *pathos* means suffering. For those of you who are actors or artists, you know that pathos is a pouring out of emotion or compassion for the suffering of others.

In science and medicine, the root *path-* is seen in terms such as *pathology,* the study of the structural changes produced by diseases (such as tumors caused by cancer and how they affect an organ). A *pathogen* is an agent that causes disease, such as a bacterium or virus. And, in this section, the focus is on *pathophysiology,* the study of the functional abnormalities that occur when a person has a disease.

For example, when a woman develops lung cancer, a pathologist looks at test results to determine the location and size of the tumor, whether the tumor is benign or malignant, and how far it has progressed. The pathologist examines the structural changes that have occurred in the lung as a result of the tumor. A pathophysiologist, on the other hand, focuses on the abnormalities in lung function that occur because of the tumor. For example, these abnormalities could limit the expandability of the lung so that the person can't take in as much oxygen as normal and/or lessen the blood's ability to oxygenate in the air sacs of the lung and thus affect other physiological functions.

Throughout this book, wherever possible, I include a section on pathophysiology to accompany normal physiology so that you can understand what happens to the body when illness strikes.

# Knowing What's Good For You

Exploring the human body is interesting, fascinating, challenging, and oh-so smart. If you understand how your body works, you're better prepared to prevent illness and improve your health. You see the wisdom in making wise lifestyle choices, and you have an easier time not only understanding what your doctor says, but why.

As you continue your journey through the human body and this book, please keep in mind that each system of the body interacts with other systems of the body. What you put in your body or do to your body has the potential to affect your entire body. Gaining the understanding takes time, but the time is well spent. My sincere hope is that this book adds to your understanding of the human body and your ability and desire to better care for yours.

# Chapter 2

# Spanning the Ages: What Your Body Does Throughout Its Life

*E*veryone is busy these days. Activities for every member of the family keep our schedules full nearly every night. The kids' homework requires time, and then there are household chores, community responsibilities, volunteer duties, religious obligations, and oh yeah, work. The list is endless. Just be thankful that you don't have to add items to your list like "Don't forget to breathe," and "Make sure your heart is beating." Your body takes care of those basic needs for you through the automatic, continuous cellular processes of metabolism. This chapter explains why and how.

## *Keeping Your Body Running: Metabolism*

*Metabolism* describes all the chemical reactions that are happening in the body. Some of the reactions, called *anabolic reactions,* create needed products. Other reactions, called *catabolic reactions,* break down products.

To keep the meanings of anabolic and catabolic clear in your mind, associate the word "catabolic" with the word "catastrophic" to remember that catabolic reactions break down products. Then you'll know that anabolic reactions create products. Your body is performing both anabolic and catabolic reactions at the same time and around the clock to keep your body alive and functioning. Even when you are sleeping, your cells are busy. You just never get to rest (until you're dead).

## Why your cells metabolize

Think of your body as a car. The engine must have gasoline and oil in order to run. The fuel system must get the gasoline into the engine. The radiator must have water to make sure that the engine doesn't overheat, and the exhaust system must eliminate hydrocarbons that can clog the engine. Every cell in your body is part of your engine because every cell converts fuel to useable energy. Your fuel system is your digestive system, which makes fuel available to your cells; instead of an exhaust system that removes pollutants, you have an excretory system. And like a car, you need water.

Your body's fuel is food. When you eat a meal, your digestive system (see Chapter 11) breaks down the food into the smallest possible pieces. Those "pieces" are the nutrients — vitamins, minerals, glucose, fatty acids, and amino acids — that your body can use as well as the waste products that are sent on to your excretory system. The nutrients are absorbed from the digestive system into the bloodstream, and the blood transports the nutrients throughout the body to all the cells. The cells then use the nutrients to fuel the reactions of metabolism. During metabolic reactions, the energy contained in the nutrients is converted to a compound called *adenosine triphosphate* (ATP), which is the form of energy used by all cells. So, nutrients are catabolized (broken down), and ATP is anabolized (created). Your cells cannot directly use "molecules of pizza" to keep you alive, but it does use the molecules of ATP created from the pizza that you eat.

## How your cells metabolize

Chapter 11 gives you the details of how the digestive system breaks down food and gets it into your bloodstream. Chapter 9 explains how the bloodstream carries nutrients around the body to every cell and waste products to the excretory system. Chapter 12 shows you how the urinary system filters the blood and removes waste from the body. But this chapter supplies the details of the reactions that your cells go through to convert the fuel to useable energy. Ready?

The reactions that convert fuel to useable energy include glycolysis, the Krebs cycle, and oxidative phosphorylation. Altogether, these reactions are referred to as *cellular respiration* (see Figure 2-1). Be sure to take it slowly through these pathways. Feel free to re-read if necessary. No one is right behind you with a case of road rage.

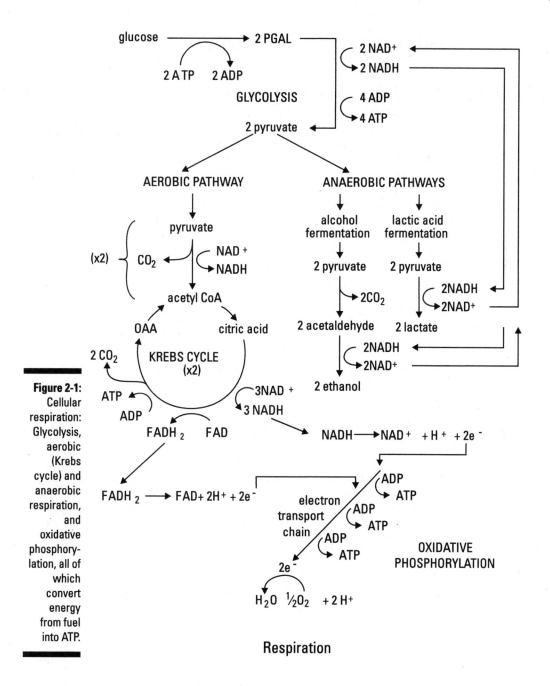

**Figure 2-1:**
Cellular
respiration:
Glycolysis,
aerobic
(Krebs
cycle) and
anaerobic
respiration,
and
oxidative
phosphory-
lation, all of
which
convert
energy
from fuel
into ATP.

Respiration

*Glycolysis,* the process that breaks down glucose, occurs in the *cytoplasm* (fluid portion) of every cell. The product of glycolysis — pyruvic acid — moves from the cytoplasm into the cellular organelle (see Chapter 3) called the *mitochondrion.* The mitochondrion is the "powerhouse" of the cell because it's where energy conversion takes place. The Krebs cycle — also called the tricarboxylic acid cycle or citric acid cycle — begins in the mitochondrion once pyruvic acid enters the mitochondrion from the cytoplasm.

At the completion of the Krebs cycle, the high-energy molecules that are created during the Krebs cycle move into the *membrane* of the mitochondrion, where they're passed down the electron transport chain. At the end of that chain, the high-energy molecules are converted to ATP, and water is released.

ATP is the "currency" of the cell. When the cell needs to use energy — such as when it needs to move substances across the cell membrane via active transport (see Chapter 3) — it "pays" with molecules of ATP. Just as you can't keep spending money without earning some money to replenish your supply, your body can't keep expending energy without taking in more fuel from which it can convert energy into ATP molecules. Converting fuel to ATP is even more crucial than earning more money, though. Without money, you would indeed be bankrupt, and that's an awful situation; but, without ATP, you die.

Look back at Figure 2-1. And keep looking back at Figure 2-1 as many times as necessary as you read through what is happening in the pathways and cycles.

### Glycolytic pathway (glycolysis)

Starting at the top of Figure 2-1, you can see that glucose — the smallest molecule that a carbohydrate can be broken into during digestion — goes through the process of glycolysis, which starts cellular respiration and uses some energy (ATP) itself. Two molecules of ATP are required to get the glycolytic pathway rolling. So although four molecules of ATP are generated during the ten steps of glycolysis (I spared you all the details), the overall amount of ATP produced is only two molecules of ATP (because two molecules of ATP are expended during the beginning of the process). In addition to the two resulting ATPs, pyruvic acid (also called pyruvate) is generated during glycolysis. And that pyruvic acid is what enters the aerobic or anaerobic pathways of the Krebs cycle and goes on to convert what started out as fuel in the form of glucose into useable energy in the form of ATP.

### Krebs cycle

Two molecules of pyruvate enter the Krebs cycle, which is called an aerobic pathway because it requires the presence of oxygen in order to occur. This cycle is a major biological pathway that occurs not only in humans but in every animal and plant.

## Digging deeper into the Krebs cycle

With the loss of water, citric acid changes to cis-aconitic acid. More water is taken in, and cis-aconitic acid becomes iso-citric acid. At this point, NAD$^+$ joins in, converting iso-citric acid to _-ketoglutarate; the reaction gives off carbon dioxide and NADH. The _-ketoglutarate converts to succinyl-coenzyme A when NAD$^+$ and coenzyme A are added. Carbon dioxide and NADH are given off in this reaction. Succinyl CoA is joined by guanosine diphosphate (GDP) and an inorganic phosphate molecule (Pi) to form succinic acid. Coenzyme A and guanosine triphosphate (GTP) are given off. Succinic acid (or succinate) is converted to fumaric acid (fumarate) when oxidized flavin-adenine dinucleotide (FAD) is added. FAD is an electron carrier like NAD+ and it is also considered to be a nonprotein enzyme. That means it helps to pass on the energy to keep the reactions moving so that the ultimate goal can be reached. FAD is reduced to FADH2 in this reaction. At this point

in the cycle, more water is added to fumarate (see why you have to take in water?), which converts the fumarate to malic acid (malate). NAD+ joins the cycle again and converts malic acid to OAA. NADH is given off. After one spin of the Krebs cycle, you have the following amounts of energy-laden molecules:

- Three molecules of NADH (reduced NAD)

- One molecule of FADH$_2$ (reduced flavin adenine dinucleotide)

- One molecule of ATP

Okay. It's pretty easy to understand that one molecule of ATP equals one molecule of ATP. But if ATP is the only energy molecule the body can use, how many ATP molecules do you get out of NADH and FADH$_2$? Read on. (Hint: NADH and FADH$_2$ are converted to ATP during oxidative phosphorylation.)

As the pyruvate enters the mitochondrion, a molecule of a compound called *nicotinamide adenine dinucleotide* (NAD$^+$) joins it. NAD$^+$ is an electron carrier (that is, it carries energy), and it gets the process moving by bringing some energy into the pathway. The NAD$^+$ provides enough energy that when it joins with pyruvate, carbon dioxide is released, and the high-energy molecule NADH is formed. The product of the overall reaction is acetyl *coenzyme A* (acetyl CoA), which is a carbohydrate molecule that puts the Krebs cycle in motion.

Cycles are circles. Products of some reactions are then used to start other reactions. An example is acetyl CoA: it is a product of the Krebs cycle, yet it also helps initiate the cycle. With the addition of water and acetyl CoA, *oxaloacetic acid* (OAA) is converted to *citric acid*. Then, a series of reactions proceed throughout the cycle. I have written them out in the following sidebar so you can see how citric acid then ends up as OAA.

### Oxidative phosphorylation (also known as the respiratory chain and the electron transport chain)

The electron carriers produced during the Krebs cycle — NADH and $FADH_2$ — are created when their oxidized partners ($NAD^+$ and FAD, respectively) become reduced. When a substance is *reduced,* it gains electrons; when it's *oxidized,* it loses electrons. So NADH and $FADH_2$ are compounds that have gained electrons, and therefore, energy. In the respiratory chain, oxidation and reduction reactions occur repeatedly as a way of transporting energy. In fact, the respiratory chain is also called the electron transport chain. At the end of the chain, oxygen accepts the electron, and water is produced.

Cells produce water as they metabolize. Some water remains in the body to regulate temperature and perform other functions. Some water is lost through sweating and exhaling. The urinary system removes some water. Because the body makes less water than it uses, you must drink water daily — 64 ounces (which equals two quarts or one-half gallon) is recommended.

As NADH and $FADH_2$ pass down the respiratory (or electron transport) chain, they lose energy as they become oxidized and reduced, oxidized and reduced, oxidized and . . . . It sounds exhausting, doesn't it? Well, their energy supplies become exhausted for a good cause. The energy that these electron carriers lose is used to add a molecule of phosphorus to adenosine *di*phosphate to make it adenosine *tri*phosphate — the coveted ATP. And ATP is the goal for converting the energy in food to energy that the cells in the body can use. For each NADH molecule that is produced in the Krebs cycle, three molecules of ATP can be generated. For each molecule of $FADH_2$ that's produced in the Krebs cycle, two molecules of ATP are made. Throughout the entire process of *aerobic cellular respiration* — glycolysis, Krebs cycle, and oxidative phosphorylation — a total of 36 ATP molecules are generated from the energy in one molecule of glucose.

### Anaerobic respiration

Sometimes oxygen isn't present, but your body still needs energy. During these rare times, a back-up system, an *anaerobic pathway* (called anaerobic because it proceeds in the absence of oxygen) exists. Lactic acid fermentation generates $NAD^+$ so that glycolysis, which results in the production of two molecules of ATP, can continue. However, if the supply of $NAD^+$ runs out, glycolysis can't occur, and ATP can't be generated.

# Transferring Energy: A Body's Place in the World

To understand why your body does what it does on a daily basis, it may help if you can develop a sense of the continuity of life — the life cycle. The *life cycle* is about energy transfer. And, you, my friend, are part of the life cycle.

The world was created with a certain amount of energy. Basically, the same energy is still here; it's just been moving throughout the world for billions of years now. Energy is within the earth and the earth's products, such as oil, coal, volcanic gases, minerals, water, and other elements. Energy is within organisms, such as bacteria, fungi, plants, and animals, and energy is in the atmosphere. It's in sunlight, oxygen, nitrogen, and so on. Energy gets transferred from the atmosphere to the earth to the organisms and then back to the earth and the atmosphere. An example of energy transfer, plants absorb nutrients and water from the soil and energy in the form of light from the sun.

Plants convert the nutrients and light energy into carbohydrates. Suppose the plant is an apple tree. Now, suppose that a deer eats one of the apples, thus taking some of the energy from the apple tree. The deer converts the carbohydrates in the apple into ATP, so it has energy to run away from a hunter. Unfortunately, the hunter wins, and the deer is killed. The hunter eats the meat of the deer, thus taking some of the energy from the deer, and the rest of the carcass is buried (returning energy to the soil) or burned (returning energy to the atmosphere). Now, the hunter has energy from the plant (the apple that the deer ate) and from the deer. As the hunter digests his meal, the fuel in the food is converted to ATP that he needs in order for his cells to function. When the hunter hikes through the woods the next day, he is returning energy to the atmosphere — heat from his body.

Eventually, when the hunter dies, all the energy he has locked up in his cells is returned to the soil and atmosphere. Other organisms use that returned energy: the daisies he pushes up, the worms in the soil, and the bacteria that consume rotting flesh. It may sound gross, but it's fundamental to life. You're an organism that harbors energy while you're alive; you return the energy you "borrowed" throughout your life to the air, water, and soil after you die so that other organisms can continue the circle of life.

The world's energy is always here. Some energy is in the organisms living on Earth, some is in the atmosphere, and some is in the soil and water. A healthy ecosystem has a good balance of energy in all those places. For more information on ecosystems and the cycles that make up the life cycle, see *Biology For Dummies* by Donna Rae Siegfried (Hungry Minds, Inc.).

# Keeping the Body Balanced: Homeostasis

*Homeostasis* describes the fine tuning that your body goes through automatically to maintain or restore balance among its systems. Anything that stresses the body — such as pain, heat or cold, infection or depleted oxygen level, and, yes, stress itself — creates imbalance. When the body's systems are imbalanced, the body's cells don't work at their optimal level. However, they try to get back to normal as quickly as possible. The body can truly heal itself.

Your body's many checks and balances ensure that it performs at its peak capability. Some of these checks include the brain's ability to continually monitor

- ✔ The amount of glucose in the blood

- ✔ Blood pressure

- ✔ Body temperature

- ✔ The pH of the blood (pH measures how acidic or basic a solution is)

All these checks are crucial to the body's proper functioning. Some of the body's balances include sweating if body temperature gets too high and the secretion of insulin if the blood glucose gets too high. The body's normal internal environment stays relatively constant (within a range of normal), even though the environment outside the body can change drastically. Normal values for important properties of the blood are

- ✔ The blood's glucose level should be about 0.1 percent. (See Chapter 8 for information on how the hormone insulin helps maintain the blood glucose level within the normal range.)

- ✔ Blood pressure is normally about 120/80 millimeters of mercury (mm Hg). (See Chapter 9 for more on blood pressure; see Chapter 12 for more on blood pressure in relationship to the kidneys.)

- ✔ The blood temperature is normally about 98.6 degrees Fahrenheit.

- ✔ The pH of the blood must be approximately 7.4. (See Chapter 9 for problems relating to pH imbalance — acidosis and alkalosis.)

The brain contains a gland called the *hypothalamus,* which lies right above the pituitary gland and is responsible for several extremely important functions. (See Chapter 8 for more on the functions of the hypothalamus and the pituitary gland.) One of the most important functions of the hypothalamus, however, is maintaining homeostasis. Throughout the body, receptors in the arteries and veins detect pressure, temperature, pH, and glucose level. As the blood flows through the blood vessels and passes over the receptors, those receptors send a signal through the nervous system to the hypothalamus. To make adjustments, the hypothalamus initiates the release of hormones that alter the blood levels or it sends a signal back through the nervous system to cause a physiologic reaction (such as shivering when you're cold).

As the blood continues to flow through the vessels, the receptors continue to send signals to the hypothalamus, which continues to monitor the blood levels. When the blood pressure, temperature, pH, or glucose level returns to normal, the hypothalamus stops sending the signal to secrete the hormone that is making adjustments. Because the *absence* of a signal to secrete a hormone is what ends the response, the process is called *negative feedback inhibition.* (The presence of something would be considered positive; therefore, the absence of something is considered negative.)

Homeostasis is an extremely important concept in physiology. Without homeostasis, your body wouldn't remain at normal levels for proper functioning. Extreme changes in temperature — like going from a house that is 80 degrees Fahrenheit to the outside, where it is 10 degrees Fahrenheit below zero would wreak havoc on your system if it were unable to adjust. Or when you develop a fever in an attempt to fight an invading organism, your system could overheat if it were not for the adjustments that occur because of homeostasis. When your body temperature is higher than 98.6 degrees F, you feel warm and may begin sweating to reduce body temperature; if it's lower than 98.6 degrees F, you feel chilly and may begin shivering to increase body temperature. Homeostasis is another important physiologic mechanism that you don't have to think about. It just happens. The body is simply amazing.

# Moving: You're Not a Tree

One of the most cherished memories that I have of my children is when I "planted" my 4-year-old son on the beach one morning in Amelia Island, Florida. We dug a hole, he stuck his feet in, and I filled in the hole and added water. He stood there immobilized and reaching for the sky, totally convinced that he was growing taller. He was happy like this for a while, until his little sister went after his just-collected shells, and he found that he couldn't protect his new-found treasures.

Plants are immobilized objects. They can't defend themselves against an animal that's going to eat it. But, then, they don't care. Plants don't think about consequences. They don't have the *fight-or-flight response*. Can you imagine if every plant was able to run from an animal that was about to take a bite? Perhaps it's a good design to have plants being stationary. There is already enough traffic.

Alas, people and other animals aren't "planted." We need to be able to run from prey. We need to go after food; human beings can't suck up nutrients and water from the soil like plants do. Okay. I know the option exists to stay planted on the couch and have food delivered to you, but you still have to get up and pay the kid at the front door. Movement *is* a necessity.

The musculoskeletal system allows voluntary movement of your limbs, head, and torso, but it also allows involuntary movements that occur inside your body: the beating of your heart, the movement of your diaphragm and rib cage to allow breathing, the squeezing of foodstuff through your digestive system (peristalsis), and many more. (See the chapters in Part III of this book for more examples of movements that occur inside your body.)

However, the bones and muscles of your body are not the only parts that move. Movement occurs even at the cellular level. The body is made of about 80 percent water, so cells are living in a mostly fluid environment, which encourages movement. But, some cells are designed to move on their own, and other cells contain structures that move. Cilia and flagella are two cellular structures that promote movement at the cellular level.

*Cilia* are tiny, tubular structures that project from the outside of a cell membrane. The purpose of cilia is to move substances over the surface of a cell. Cells that contain cilia are found in the upper respiratory tract. The cells in the nose, sinuses, and throat secrete mucus to trap debris; the cilia act like brooms and move the particles to the throat so that the lungs remain clear.

*Flagella* also are microtubular structures attached to cell membranes. However, the movement of flagella is a bit different than that of cilia. Flagella move in a whip-like fashion, which helps cells progress through the fluid environment of the body. Flagella are found in sperm cells, and they help move sperm through the fluid secreted by female reproductive organs so that the sperm can get to the egg (see Chapter 14).

# Continuing the Species: Reproduction

The need for reproduction should seem pretty obvious — if humans didn't mate, there would be no more humans. The same is true of any species; reproduction is essential to the circle of life.

Some organisms reproduce by asexual reproduction, which basically is the same as cell division (see next section). When the organism starts to wear out, it divides into two new organisms. But, most animals, including humans, are lucky enough to have sexual reproduction. Sexual reproduction requires two members of a species — no pun intended — rather than one old cell.

You might think that sexual reproduction in humans starts when a man and a woman actually mate. The act of intercourse is how genetic material is transferred, certainly, but it's not where the whole process begins. The starting point actually is at the cellular level. In females, it starts before a woman is even born. In males, it happens continually. "It" is the production of gametes: eggs and sperm. For the details of this intriguing process, head to Chapter 14.

# Growing: Replacing Cells and Developing Throughout Life

Growth may seem as simple as growing taller, but did you ever think about *how* you get taller? Your bones get longer, your muscles stretch, and organs enlarge. And because organs, bones, and muscles are all made of tissues, which are made of cells, more cells must be produced for growth to occur.

Cells are produced by dividing, and the process of cell division is methodical. There are two types of cell division:

- **Meiosis:** When sperm and eggs are produced in preparation for reproduction, meiosis occurs in the testes or ovaries. (See Chapter 14 for more on human reproduction.)

- **Mitosis:** When cells need to undergo repair or growth, mitosis occurs in every "regular" (somatic) cell — meaning non-sex cell — of the body.

During mitosis, one cell divides into two cells. The cell that is about to divide is the *mother cell;* the two cells that are produced are *daughter cells.* To understand what happens during mitosis, you need a little background in genetics. The reason for needing a grounding in genetics is that for one cell to divide into two cells, the genetic material inside the nucleus of the mother cell must be duplicated so that each daughter cell gets a complete set of genetic information. You can find the details of mitosis in the next section.

# Genetic material: DNA, chromosomes, and genes

Inside each cell nucleus are *chromosomes*. The chromosomes are made up of *chromatin,* which is made of protein and *deoxyribonucleic acid* strands. Deoxyribonucleic acid is *DNA* — the genetic material that's in the shape of a twisted ladder called *a double helix.* Along the strands of DNA lie molecules called nucleotides. Each nucleotide contains

- Deoxyribose (a sugar with five carbon molecules)
- A phosphate group
- A nitrogenous base

The deoxyribose and phosphate molecules attach to the nitrogenous base to form the nucleotide. DNA has four different nitrogen-containing molecules that it can use as the base for a nucleotide:

- Adenine (A)
- Cytosine (C)
- Guanine (G)
- Thymine (T)

The nitrogenous bases (and therefore the nucleotides) can be and are different throughout the long chain of DNA. Groups of nucleotides make up *genes.* The nucleotides are nitrogen-containing molecules that form the "rungs" of the "twisted ladder" called DNA. A gene can be made up of a few "rungs" or thousands of "rungs;" the variety is extensive (which accounts for the variety of characteristics among living things). Nucleotides and genes aren't separate from DNA; they kind of co-exist together like a set of nesting boxes. The largest box is the DNA, and it holds several smaller boxes — the genes. But the genes are made up of many still smaller boxes — the nucleotides.

*Genes* direct your body; they contain the blueprint for what you look like, how you age, what diseases you're susceptible to, your strengths, your weaknesses, in short, everything that makes you, you. See, genes control the production of *amino acids* that join together to form proteins, and proteins are found in everything: cell membranes; tissues of the skin, bones, muscles, organs, hormones, and enzymes. Proteins direct growth and development as well as many daily functions (such as homeostasis). And genes on the DNA serve as the blueprint for what amino acids form what proteins, where the proteins are produced, and when they are produced.

So you may have a gene for dark brown skin, but that gene doesn't produce your dark skin by itself. It contains the instructions for what amino acids your melanocytes (pigment-producing cells of the skin) "pull off the shelf" and in what order to assemble them. The order of the amino acids determines the protein that's produced. That protein directs the amount of pigment produced by your melanocytes.

Genes are responsible for your traits — the outward expression of what's in your genetic makeup. If your genes dictate that you'll grow to six feet, they cause bone and tissue growth to occur until you reach that height. Your height is then the outward expression of your genes for height. If your genes say that your eyes will be blue, the iris (the colored part) of your eye produces proteins that allow the blue color — the outward expression of your genes for eye color. And if your genes include the trait for curly hair, your hair follicles produce hair cells in a wavy pattern. Your ringlets are the outward expression of your genes for hair texture.

In humans, thousands of genes reside on a total of 46 chromosomes in the "regular" cells of your body. Of those 46, 23 chromosomes come from your father, and 23 come from your mother. Each sex cell (gamete) such as a sperm or an egg, which are single cells, contains 23 chromosomes, so that when a sperm and egg join, they result in a new cell containing the full 46 chromosomes. But each "regular" cell in your body — and by "regular" I mean nonsex cells (also called *somatic cells*) that make up the tissues and organs, not the sperm or eggs that you have in your reproductive system — contains a full set of chromosomes from both parents. Normally, in a nondividing cell, the chromosomes exist as a single strand of DNA. However, prior to the separation of dividing cells, the DNA must replicate (copy itself) so that a full set of chromosomes is in the original cell as well as in the new cells it's producing.

## *Making more DNA and chromosomes*

The purpose of *DNA replication* (see Figure 2-2), the process that occurs while chromosomes are duplicating, is to produce copies of both strands of DNA. The reason that both strands need to be copied is because the copies are used as the template or blueprint from which all the hormones and substances needed to run your body are made. All the cellular products and processes are controlled by the genes that are on the chromosomes that are made up of your DNA. So when a cell divides into new cells (this happens daily) or when you are mating to form a new person (this happens periodically), that genetic material needs to go along. The genetic information is needed to direct the development and growth of the organism throughout the organism's life.

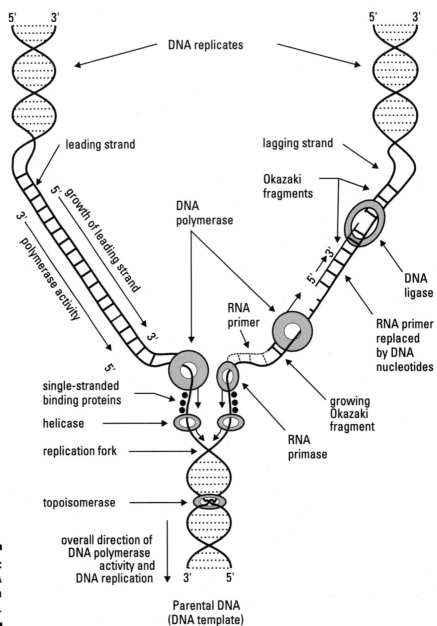

**Figure 2-2:**
The DNA
replication
process.

During DNA replication, the "twisted ladder" of the DNA double helix must untwist and "unzip" so that the "rungs" of the ladder are split apart. This splitting is initiated by the enzyme *helicase*; the result is a Y-shaped DNA molecule, with one nucleotide on the left branch of the Y and one nucleotide on the other side (refer to Figure 2-2). Each side of the original DNA strand becomes a *template strand*. A template is a pattern or a mold that's followed to create something new. In genetics, the template strand serves as the pattern for the new *complementary strand*.

Complementary strands form on each template strand as the enzyme *DNA polymerase* adds a nucleotide based on the order of nitrogenous bases on the template strand. The four nitrogenous bases of DNA pair up: A pairs with T, and G pairs with C. So wherever the original template strand has an A, the complementary strand gets a T added by the DNA polymerase. This process occurs a little at a time along a strand of DNA. The entire DNA strand does not unravel and split apart all at once. When the top part of the helix is open, the original DNA strand looks like a Y (refer to Figure 2-2). This partly open/partly closed area where replication is happening is the *replication fork*.

The order of the bases is important because the order of bases delineate the genes, and the genes dictate what amino acids are produced, and the amino acids determine which proteins are produced, and proteins are needed in every cell of your body. Proteins make up cell structures themselves, as well as enzymes that initiate cellular processes that keep you alive.

When you look at Figure 2-2, you'll see the numbers 5' and 3' (read "five prime" and "three prime"). These numbers indicate the direction in which DNA replication is occurring: The template strand is read in the 3' to 5' direction. The bases that are complementary to the original template strand are added opposite the template strand, so the "new" complementary strand "grows" in the 5' to 3' direction.

Mistakes can happen as the DNA polymerase reads the template strand and puts the corresponding nucleotide on the complementary strand. If the template strand has an A, and DNA polymerase puts a C on the complementary strand, the genetic information is inaccurate. Fortunately, Mother Nature thought of everything. If a mistake occurs, a proofreading function in the nucleus detects erroneous bases. The bad bases are eliminated, and the DNA polymerase tries again. If there is still a mistake, *mismatch repair enzymes* in the nucleus attempt to get the order of the bases just as they were in the original DNA strand.

However, sometimes errors (called *mutations*) remain. Mutations sound serious, but in fact, they contribute to the development and evolution of a species because they allow changes to happen. Mutations usually result from radiation, such as ultraviolet light and x-rays, or certain chemicals. Three major types of mutations affect the order of nucleotides on a DNA strand, and thus the bases that make up a gene.

- ✔ **Insertions** occur when an extra nucleotide is added to the complementary strand. This throws off the reading of the genetic code on the new strand of DNA by one or hundreds of bases. If the genes cannot be interpreted correctly, the wrong amino acids can be produced. Therefore, the wrong proteins can be produced, and this can have devastating results. This type of mutation can result in *Huntington's disease,* which causes degeneration of the nervous system.

- ✔ **Deletions** occur if a nucleotide is read from the template strand, but the complementary base is not added to the new strand (that is, a base is left out of the complementary strand). This type of mutation causes *cystic fibrosis* and *Duchenne muscular dystrophy* — two serious diseases.

- ✔ **Substitutions** are the least serious type of mutation. One base is wrong in a group of many bases that form a gene. Because just one base is off, these errors are *point mutations.* Most often, this type of mutation causes no ill effect; if so, the error is referred to as a *silent mutation.*

DNA replication occurs just prior to the beginning of mitosis during *interphase,* which is not a part of mitosis. As its name suggests, it's an intermediate phase that the daughter cells remain in until they're ready to act as mother cells and go through the stages of mitosis (see Figure 2-3) to divide.

The steps of mitosis are as follows (Hint: Think of cells like little globes):

1. **Prophase:** In this first stage of mitosis, the single *chromatids* (a strand of DNA) that were present during interphase thicken and coil to become chromosomes. Chromosomes aren't notorious for their ability to move; so while the chromosomes are beefing up, cellular structures called spindle fibers move to the poles of the cells, where they spawn more spindle fibers that help move chromosomes around the cell.

   The chromosomes are able to move around the cell, rather than just around the nucleus at this time, because while the chromosomes were thickening and the spindle fibers were creating spindles, the membrane around the nucleus was disintegrating. The *centrioles* (another of the cellular organelles) get to swing into action in this stage. Centrioles attach to the spindle fibers, which move to the poles of the cell, and spindles run from one end of the cell to the other.

2. **Metaphase:** By the time the cell reaches metaphase, the nucleus, making room for the chromosomes to get into position, is completely gone.

During metaphase, the chromosomes move toward the cell's center, forming a perfect row along the cell's equivalent to the Earth's equator. At this point, 92 chromatids still make up 46 chromosomes.

3. **Anaphase:** During anaphase, the chromatids strike out on their own. The chromatids move along the spindles to opposite poles of the cell; 46 to one pole, 46 to the other. Following this movement, the chromosomes are referred to as *daughter chromosomes,* but the cell isn't quite ready to divide yet.

4. **Telophase:** Okay. Now the daughter chromosomes are ready for their own cells. During telophase, the set of chromosomes at one pole is identical to the set of chromosomes at the opposite pole. A fresh nuclear membrane surrounds each set of chromosomes. The spindles dissolve, which frees the daughter chromosomes.

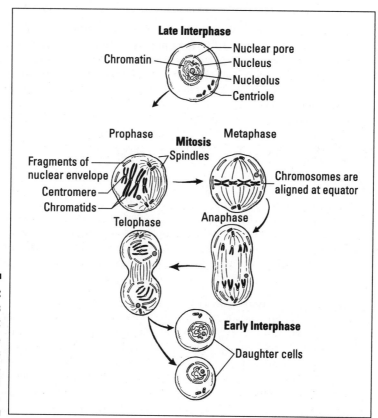

**Figure 2-3:**
The stages of mitosis: Prophase, metaphase, anaphase, and telophase.

At this point, mitosis is technically over, and early interphase begins (remember, this is a cycle that's going on here, so if interphase occurs at 12 o'clock" you've "gone around the clock" so to speak). However, the cell still has to actually split apart into two cells — a process called *cytokinesis*. It happens when the center of the mother cell indents — the indentation is called a *furrow* (as is the line between your eyebrows when you look confused or concerned). The furrow squeezes the cell membrane into the cytoplasm until two separate cells are formed. The moment of separation is *cell cleavage*. Now the two daughter cells can live out their time in late interphase, until they need to divide.

Cell division occurs when new tissue is needed to heal a wound or regenerate part of the body. (Yes, you can regenerate some parts of your body, such as your liver.) And it occurs as a part of daily life. On any given day, your body is replacing cells that are wearing out. For example, each blood cell has a life span of 120 days. But you don't get a full set of new blood cells every 120 days. Your body replaces some of them every day. Also, you're constantly growing hair and nails. (Your nose and feet also continue to grow throughout your life, by the way.) And cells are constantly metabolizing to provide you with the ATP that keeps you alive. So replacement of cells is as essential to life as creating new ones. Consider yourself lucky though. For some organisms, like fungus, cell division is their form of reproduction. At least you have a different means for that!

## Building blocks that build you

After the DNA replicates, and the fresh, new complementary strand is completed, the cell needs to move the strand to the next step in the assembly line of protein production. DNA replication occurs in the cell nucleus, but the amino acid building blocks that form proteins are put together outside of the nucleus in the cell's cytoplasm. In this section, you find out how the instructions in the new DNA strand get to the cytoplasm so that the proteins can be made.

You've heard about DNA, now let me introduce you to DNA's cousin, RNA, an acronym that stands for *ribonucleic acid* (no *deoxy-* in front of its name). RNA is similar to DNA except that RNA has only one strand instead of two; the sugar molecule it contains is ribose instead of deoxyribose; and instead of thymine (T), it has the nitrogenous base *uracil* (U). So in RNA, adenine (A) pairs with U. You need to know about RNA because RNA is the molecule that transports the genetic information out of the nucleus and puts together the proteins. Three forms of RNA perform specific tasks in this process:

1. **Messenger RNA (mRNA)** carries the message from the DNA out of the nucleus to the ribosome, the organelle where protein synthesis occurs.

2. **Ribosomal RNA (rRNA)** translates the DNA message three nucleotides at a time.

3. **Transfer RNA (tRNA)** brings the appropriate amino acid to the site of protein production.

During the process of *transcription*, the complementary strand of DNA that was just produced from the original template strand of DNA now serves as the template (see Figure 2-4) for the production of a strand of messenger RNA (mRNA). The complementary DNA strand has nucleotide bases that are opposite the original template strand's order of bases. The pairs of A-T and G-C are matched. But now, the complementary strand is "read," and a strand of mRNA is produced. So where the original DNA's template strand had the nucleotides T-G-G-T, it's complementary strand has the nucleotides A-C-C-A, and the mRNA that forms has the nucleotides U-G-G-U.

*Remember:* RNA molecules use uracil instead of thymine.

This may seem like an extra step — why create a strand of RNA that's pretty close to the order of bases in the original DNA? Well, the original strand must be copied so that new cells or new organisms can be created. And the complementary strand of DNA that's created from the original can't leave the nucleus. So the message contained in the complementary DNA strand must be transcribed into RNA (like transcribing English letters into Greek letters) in order for it to be carried outside the nucleus.

Certain sequences of nucleotides on the mRNA strand indicate where transcription should begin, and certain sequences indicate where transcription should stop. And, along the strand of mRNA there are sequences that go untranscribed. These areas are like fillers; they don't contain any truly useable information. So, just like frames of a filmstrip, those pieces — called *introns* (for <u>in</u>tervening sequences) — are spliced out. The translated sequences — called *exons* (expressed sequences) — are joined together. Now, the strand of mRNA contains only nucleotide bases that will be translated.

When the mRNA molecule leaves the nucleus and enters the cytoplasm, it carries the genetic information to a *ribosome* (a cellular organelle). In the ribosome, the process of *translation* takes place. (So now, if English letters were transcribed into "Greek letters," the Greek letters are translated into Greek words.) The mRNA strand passes through the ribosome and, as it does, the ribosome, reading three nucleotides at a time, translates the "words." The group of three nucleotides is a *codon,* and each codon specifies an amino acid.

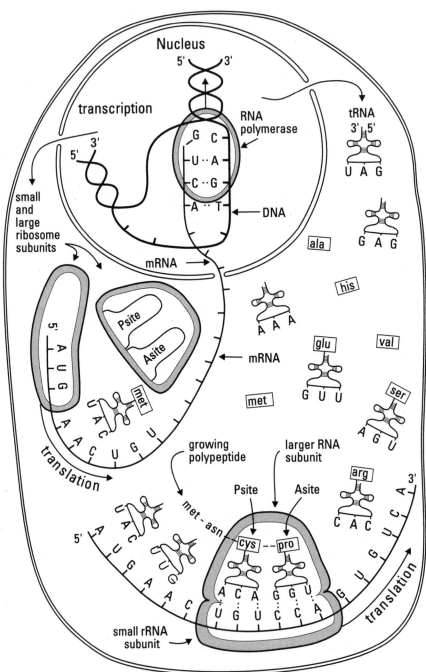

**Figure 2-4:**
The process
of protein
synthesis:
Transcription
in the
nucleus,
translation
in the
ribosomes.

Protein Synthesis

The *genetic code* — which, amazingly, is the same for all living things — is the language that bridges the gap between genes and amino acids. If the group of three nucleotides *(codon)* is UCG, the amino acid that is being specified by the genetic material is *tryptophan*. (Twenty amino acids exist in nature, and the genetic code contains 64 different codons; so obviously most amino acids are coded for by more than one codon.)

As each codon is read, a molecule of tRNA brings the appropriate amino acid molecule to the ribosome. Picture an amino acid as a building block and the molecule of tRNA as a warehouse worker. The codon is the written instruction as to what building block the warehouse worker should retrieve from the shelves. When the warehouse worker brings in the block, the next warehouse worker gets his instructions to get another block. Block by block, amino acid by amino acid, a protein is built. The protein may need a little tweaking to become fully functional, but it doesn't take long. After the genetic instructions are read, and the amino acids are put in place, the organism soon uses the new protein.

And like I said, your body continually needs new proteins to build new cells and tissues, repair injuries, fight off microbes, and create necessary enzymes and hormones to keep your body's systems running smoothly.

# Chapter 3

# Forming Your Foundation

· · · · · · · · · · · · · · · · · · · · · · · · · · · · · · · · · · · · · · · · · · ·

· · · · · · · · · · · · · · · · · · · · · · · · · · · · · · · · · · · · · · · · · · ·

T his chapter serves to make sure that you have a good understanding of your body's basic processes. If you want to delve into the larger systems of the body, see Part II. In this chapter, you see why each cell in your body is a miniature organism that performs the processes, such as converting nutrients into energy and protein, which you as a large organism perform. You'll take a look at how cells are put together to form tissues that go on to make up all the rest of the parts of your body: bones, muscles, blood vessels, nerves, and organs. Because cells and tissues comprise every part of your body, they can be considered the foundation of the body. Now, take a look at how that foundation forms.

# Selling You on Cells (No Phones, Towers, or Metal Bars)

The word *cell* can describe several different things: lodging in a prison, telecommunications technology, and even parts of a table in a word processing program. But, regarding anatomy and physiology, a cell is a microscopic unit that makes a living thing a, well, living thing. Before I can convince you that cells are the fundamental unit of life, I have to describe a cell's components and make sure that you know what a cell is and does. These guys may be little, but they're important — rather, essential — to life.

# You animal, you!

Every living thing has cells; plants, animals, and fungi are the three main groups of living things. Some organisms may be just one cell in size, whereas large organisms (such as humans) contain trillions of cells. Plant cells have rigid, fibrous cell walls, and they contain chlorophyll, a pigment that makes leaves look green. But animals — and humans are indeed animals — have cells with no cell wall or chlorophyll. Animal cells are enclosed by cell membranes, not cell walls. All animals have cells with the same basic components; the differences between animals cells lies in the genetic material inside the cell (see Chapter 2).

# Seeing the inside of a cell

A *cell* is a sac of gel-like material within a cell membrane. (To see a cell's structure, look at Figure 3-1.) The gel-like material is *cytoplasm.* (Sometimes the gel-like material is *plasma,* and the cell membrane is a *plasma membrane.* I prefer the term cytoplasm, so that's what you'll see in this book.) The cytoplasm moves around inside the membrane and pushes against the cell membrane to give it shape, much like water supports the sides of an aboveground pool. The cell membrane keeps the cytoplasm from leaking out. The cell membrane also is choosy about what can get into the cell.

## Cell structure

Embedded in the cytoplasm, *organelles* (little organs) are the cellular components inside the cell where the cellular processes take place. The organelles act a little like factories — each organelle is responsible for producing a certain product(s) that's used elsewhere in the cell or body. The most important organelles of animal cells (including yours) are given in Table 3-1.

| Table 3-1 | Organelles of Animal Cells (That includes Humans) |
|---|---|
| **Organelle** | **Function** |
| Nucleus | Controls the cell; houses the genetic material |
| Mitochondrion | Cell "powerhouse;" converts food nutrients, such as glucose, to a fuel that the cells of the body can use |
| Endoplasmic reticulum | Plays an important role in making proteins and shuttling cellular products; also involved in metabolizing fats |

| Organelle | Function |
|-----------|----------|
| Golgi apparatus | "Packages" cellular products in sacs called *vesicles* so that the products can cross the cell membrane to exit the cell |
| Vacuoles | Spaces in the cytoplasm that sometimes serve to carry materials to the cell membrane for discharge to the outside of the cell |
| Lysosomes | Contain digestive enzymes that break down harmful cell products and waste materials and then force them out of the cell |

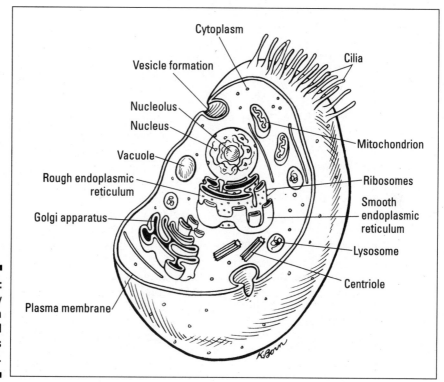

**Figure 3-1:**
A cutaway view of a basic animal cell and its organelles.

## Cell activity

As you can probably tell from the functions of organelles listed in Table 3-1, a cell is a pretty busy place. So when I tell you that your body contains several trillion cells, it may make sense to take a nap once in a while.

Here's a quick overview of the most important cellular functions, structures, activities, and traits:

- ✔ **Respiration:** The process in which the energy in food (measurable in calories) is converted to *adenosine triphosphate* (ATP) — the fuel that every cell in the body uses to provide the energy for cellular reactions that create products (a process called *anabolism*) or break down products (a process called *catabolism*); the sum total of all cellular reactions going on in the body is *metabolism*. If ATP is created in the presence of oxygen (acquired through breathing), the process of respiration is *aerobic respiration*. If oxygen is *not* used, the process is *anaerobic respiration*. For the details of these processes, go to Chapter 2.

- ✔ **Selective permeability:** The cell membrane (also called the *plasma membrane*; refer to Figure 3-1) is a bit picky about what gets into and out of the cell it surrounds. The cell isn't totally impermeable; if it were, nutrients and oxygen couldn't get into the cell. But neither are cells totally permeable because if they were, toxins and waste products could get in and damage cells quite easily. Rather, cell membranes let some things in and keep some things out: They're selective about what gets admitted. This characteristic is *selective permeability*. Cells are like exclusive nightclubs. Molecules that need to get inside better be dressed just right, or the cell membrane bounces them right back out into the bloodstream.

  The *fluid-mosaic model* (see Figure 3-2) is usually used to describe how molecules are transported across the cell membrane. In this model, the word "fluid" describes the membrane's flexibility; the word "mosaic" pertains to the fact that the membrane has large proteins and other substances embedded in it. The two layers in a cell membrane are both made up of *lipids* (fats); therefore, the construction is referred to as the *phospholipid bilayer*. The outside of both the top and bottom layers has water-loving *(hydrophilic)* heads; between the two layers are the water-fearing *(hydrophobic)* tails. Cell cytoplasm is watery, as is the *matrix* — the ground material in which cells lie. Therefore, the water-loving layers surround the cytoplasm and function well inside the body, and the hydrophobic tails are protected so they don't have to deal with the big, bad water.

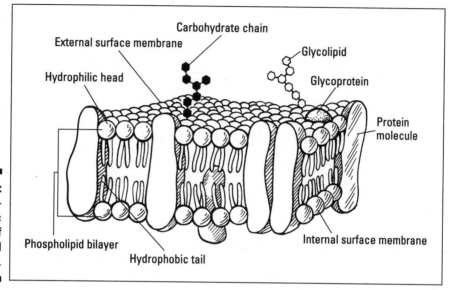

**Figure 3-2:**
The fluid-
mosaic
model of
the cell
membrane.

Carbohydrate chain

External surface membrane

Glycolipid

Hydrophilic head

Glycoprotein

Protein
molecule

Phospholipid bilayer

Internal surface membrane

Hydrophobic tail

✔ **The cell cycle:** The cell goes through periods of growth and rest called the *cell cycle* as it reproduces. If cells didn't reproduce, the organism they make up would eventually die. (See Chapter 2 for more on the cell cycle.)

✔ **Genes and proteins:** Each cell's nucleus holds a complete set of your genes, which are dotted along the strands of DNA contained in your chromosomes. When a cell must create a product needed for a process in one of your body's systems, such as creating a protein used in the hormone insulin to control the glucose level in your bloodstream, part of the genetic material must be read, matched, and translated. These processes allow the amino acids that form a protein to be produced. Your cells are constantly forming different proteins for various uses. The proteins must be created and then exported out of the cell. The processes involved in copying DNA and creating new proteins are described in detail in Chapter 2.

✔ **Autodigestion:** Old, worn-out cell parts need to be removed from the cells; if they aren't, they can become sources of toxins or severe energy drains. *Lysosomes* are the cellular organelles that do the dirty work of cleaning house. When organelles, such as mitochondria, have become unable to generate enough energy within the cell, the lysosomes release enzymes that break down the mitochondria.

So because a substance in the cell (the lysosomal enzymes) destroys another substance in the cell (the old mitochondria) through a digestive action, the process is called *autodigestion* (*auto-* means self, such as in autobiography). Molecules that can be recovered from the mitochondria are recycled so they can be used elsewhere in that cell or in another cell. (Mother Nature hates to waste things.) Then any leftover waste products are excreted from the cell. *Excretion* is the process of eliminating waste; waste is removed from every cell as well as from the body as a whole.

✔ **Cell membrane transport:** Substances such as hormones (see Chapter 8) are produced in one cell but have their effect in another cell. Hormones travel through the bloodstream to get to the cell that they're targeting, but then they need to get inside the cell to get their job done. Other substances (such as nutrients and oxygen) also need to pass through cell membranes; sometimes the substance gets transported across a cell membrane, other times it crosses the membrane by itself.

Substances pass through cell membranes via *channels*. Specific proteins create channels in the membrane. If the substance trying to get into the cell fits into another protein molecule, called a *receptor molecule,* entry is granted. This process is much like putting a key into a lock. If the pattern of the key matches the pattern inside of the lock, you can open the door. If not, you're left outside to try another door or another key. The receptor molecule ensures that the right substance works on the right cell.

Movement through a cell membrane can be *active,* which requires energy, or *passive,* which occurs based on the conditions inside and outside the cell. The differences between these types of cell membrane transport is analogous to having a door held open for you by a uniformed guard (active transport) or walking through an automatic door all by yourself (passive).

Some molecules require *active transport* with assistance from *carrier molecules* — protein molecules embedded in the cell membrane that act as carriers. The substance expends some energy molecules in order to attach to these *carrier molecules,* which then transport the substance across the cell membrane.

Substances are sometimes small enough that they can pass in or out of the cell with no energy output. The cell barely notices. *Passive transport* is like a timid cat going in and out of the house through a little hole in the back door whenever the mood strikes. Passive transport can happen in any of three ways:

- **Diffusion:** Substances move from an area where they are highly concentrated to an area where the molecules are less concentrated. The substances spread themselves across the membrane to even out the areas of concentration.

- **Osmosis:** This term is used when talking about water molecules diffusing across a selectively permeable membrane. The membrane allows water to pass through it, but keeps out the solutes dissolved in the water. As with diffusion, osmosis occurs from an area of high concentration to an area of low concentration. The difference between regular diffusion and osmosis, however, is that the area of concentration being considered is the concentration of water *(solvent)* rather than solute. Therefore, osmosis occurs from an area of high concentration of water molecules to an area of low concentration of water molecules.

  Osmosis takes into consideration the concentration of particles *(solutes)* in the water *(solvent)*. An *isotonic* solution is one that has equal concentrations of solutes and solvent. A *hypotonic* solution is one that has a lower concentration of particles (and more water) than an isotonic solution. A *hypertonic* solution has a higher concentration of solutes (and less water) than an isotonic solution. The higher the concentration of solutes in a solution, the lower the concentration of water, which increases *osmotic pressure* — the pressure at which movement of water across a membrane stops. The water balance inside and outside of a cell must be stable in order for the cell to function at a normal level. Homeostasis (see Chapter 8) helps to keep the water balance stable.

- **Filtration:** This form of passive transport occurs during capillary exchange. (Capillaries are the smallest blood vessels — they bridge arterioles and venules; see Chapter 9.) Capillaries are only as thick as one cell, so substances in tissue fluid (such as carbon dioxide and water) can move right into the capillary, and substances in the capillary (such as glucose or oxygen) can get into tissue fluid. However, movement doesn't just happen. The difference between osmotic pressure and blood pressure determines which direction substances move, and the capillary membrane acts as a filter.

  The blood pressure in the capillaries is higher at the arterial end of the capillary and lowest at the venous end of the capillary. The osmotic pressure — the pressure at which movement of water across a membrane stops — stays the same. So at the arterial end of a capillary, the blood pressure is higher than the osmotic pressure; thus a net blood pressure pushes small substances (such as water molecules, oxygen, or glucose) through the capillary membrane and into the tissue fluid. (From the tissue fluid, these substances diffuse into cells.) However, at the venous end of a capillary, the blood pressure is lower than the osmotic pressure; so a net osmotic pressure pushes waste products into the tissue fluid and pulls water into the capillary from the tissue fluid.

# Organizing Cells Into Tissues

Cell upon cell upon cell upon cell repeated over and over again results in a tissue. Tissues are formed from groups of cells that perform the same functions. So when huge numbers of similar cells get together, they form a tissue, just like when huge amounts of similarly functioning tissue aggregate into an organ, which performs a specific function.

Here's an example: Cells that secrete products, such as the juices of the digestive system, get together to form a tissue. Other cells and tissues join, and eventually an organ is created. The organ, such as the stomach, is part of a larger organ system — the digestive system — that controls when the cells in the tissues make and secrete the juices. All the levels of the body work together though. This section shows you how tissues are formed from cells and what different types of tissues — epithelial, connective, muscular, and nervous — do for you.

## Continuing with epithelial tissue: Skin

The interesting characteristic about skin (the *epithelium*) is that it's a continuous sheet that covers the entire surface of the body in addition to lining the body's cavities (see Chapter 6). Although the epithelium is a continuous layer, its specialized functions are in different areas. As skin, the epithelium is protective in the following ways:

✔ Keeps the body from losing moisture and drying out

✔ Protects against injury to internal structures

✔ Helps guard against bacterial invasion

The epithelium also lines and protects the stomach, which produces enzymes and acids that would destroy the stomach if it weren't for the epithelium's mucus secretions. The epithelium that lines the nose contains hairy projections called *cilia,* which trap dirt, dust, and other particles to keep them from getting down into your lungs.

The three types of epithelial cells are displayed in Table 3-2.

| Table 3-2 | Epithelial Cell Types | |
|---|---|---|
| *Type of Epithelial Cell* | *Description* | *Location in Body* |
| **Columnar** | Shaped like columns; the cell nucleus is generally at the bottom of the cell. | Found in lining of digestive tract |

| Type of Epithelial Cell | Description | Location in Body |
| --- | --- | --- |
| Cuboidal | Cells are shaped like cubes. | Found in small tubes of the kidney |
| Squamous | Cells are flat. Squamous means scaly. | Found in lining of lungs and blood vessels |

And the fun doesn't stop with just the three types of tissues. A few adjectives can be thrown in front of the three types, which help create several variations. (Aren't you excited?!) Epithelial tissue can be described in three more ways:

- **Simple tissues** are formed from single layers of cells.
- **Stratified tissues** have sheets of cells layered on top of each other.
- **Pseudostratified** means that the epithelium looks layered but does not form true layers.

So using these adjectives, you can put the words together to make *stratified squamous epithelium, simple cuboidal epithelium,* and *pseudostratified columnar epithelium.* And your body does make them. Stratified squamous epithelium is in the outer layer of your skin; pseudostratified columnar epithelium is found in the glands and organs that secrete products, such as in the digestive tract. Add in the adjective *ciliated,* meaning "having cilia," and you get *pseudostratified ciliated columnar epithelium* (whew!): this type of tissue is found in the lining of the nose and trachea (windpipe), and the product it secretes is mucus. Several other combinations exist, but you get the idea.

## Connecting with connective tissue

Connective tissue is a difficult tissue to visualize because it has so many functions, and thus, many different forms. In some parts of the body, such as the bones, connective tissue is supportive or provides protection. In other parts of the body, connective tissue fills in spaces and stores fat to provide the body with what amounts to "shock absorbers." Connective tissue also produces blood cells.

Generally, connective tissue is made up of cells that are spaced far apart with a fluid or gel-like matrix. A matrix is a type of ground material or base. Think of a toy train layout covered in brown moss. The brown moss, as a matrix, is underneath and between every piece of track, every building, every person. The track, buildings, and people are like cells in a matrix of brown moss.

Connective tissue matrix can contain three types of fibers:

- **White** fibers contain the strong and stretchy protein called *collagen*.

- **Yellow** fibers contains the more stretchy (elastic) but weaker protein called *elastin*.

- **Reticular** fibers are really thin, extremely branched fibers that provide support.

And the two main types of connective tissue are *loose* and *fibrous*.

### Loose connective tissue

Contrary to how its name sounds, *loose connective tissue* holds structures together. Most commonly, loose connective tissue is what holds an epithelium to the body part underneath it. For example, loose connective tissue holds the outer layer of skin to the underlying muscle tissue. When you peel the skin off a chicken breast, you're ripping apart the loose connective tissue holding the skin layer to the breast muscle. Also, loose connective tissue joins the epithelium lining the innermost skin layer covering your abdomen to the intestines and abdominal organs behind it.

Loose connective tissue is made up of cells called *fibroblasts,* which are large and shaped like stars. In loose connective tissue, fibroblasts are spaced far apart, and the matrix spread among the fibroblasts contains both collagen and elastin fibers.

The two types of loose connective tissue are

- **Adipose tissue:** More commonly known as *fat,* adipose tissue has fibroblastlike adipose cells that get larger as they fill with lipids (fat), eventually limiting the collagen and elastin fibers within the matrix. The ability of adipose cells to store fat has advantages. Fat insulates the body, protects the internal organs, and provides a source of energy when needed. However, if your adipose cells are excessively large, if you know what I mean, you've got a problem that can affect your health. Adipose tissue between your skin and abdominal organs is certainly obvious; the resulting fat belly is a telltale sign.

- **Lymphoid tissue:** Found in the lymph nodes, spleen, thymus gland, and red bone marrow — all of which are places in the body involved in immune functions (see Chapter 13), lymphoid tissue has cells like fibroblasts except they're called *reticular cells.* And the matrix of lymphoid tissue is filled with reticular fibers instead of collagen and elastin. Reticular fibers are those thin, branched fibers that form networks. Think of delicate lace.

### Fibrous connective tissue

*Fibrous connective tissue* also holds body parts together, but its structure is a bit more rigid than loose connective tissue. The fibroblasts of fibrous connective tissue are packed tightly together, and the matrix contains collagen fibers that run parallel to each other. Fibrous connective tissue is found in *ligaments,* which attach one bone to another at a joint, and in *tendons,* which attach muscles to bones.

*Cartilage* also is made of fibrous connective tissue, but it's stronger than ligaments and tendons because its matrix is more solid. However, it's not as solid as bone; cartilage matrix flexes slightly. The problem with cartilage, though, is that if it flexes too far and breaks, it has no blood supply. Because of the lack of blood supply, damaged cartilage heals extremely slowly.

Three types of cartilage exist based on the type of fiber found in the matrix of the cartilage:

- **Elastic cartilage** has many collagen fibers in addition to many elastin fibers. This type of cartilage is flexible and rarely "breaks." One place in the body that elastic cartilage is found is in the outer ear.

- **Fibrous cartilage** has mostly collagen fibers. It's strong and absorbs shock, so it's found between the vertebrae in the backbone and in the knee joint. Fibrous cartilage serves to reduce friction between joints.

- **Hyaline cartilage** contains only collagen fibers. It's strong, and is the most common type of cartilage found in the body. Hyaline cartilage looks smooth, white, and opaque. (Think about the connection between a chicken breast and the rib meat). In humans, hyaline cartilage is found in your nose, in the rings that support and protect your trachea (windpipe), and at the ends of long bones (such as legs or arms) and ribs. Fetal skeletons form from hyaline cartilage, which is later replaced with bone.

I'm sure you already know that bones are strong — bones are made of the strongest fibrous connective tissue there is. The matrix of bone is extremely hard because it contains mineral salts mixed in with protein fibers. Calcium is the most prevalent of the mineral salts found in bone, which is why you need to keep up your supply of calcium to keep your bones strong. Turn to Chapter 4 to find out what can happen to your bones if you're calcium deficient.

## Amassing info on muscle tissue

Your body contains three types of muscle tissue: cardiac, smooth, and skeletal. Those muscle tissues are made up of *muscle fibers.* The muscle fibers contain many *myofibrils,* which are the parts of the fiber that actually contract.

The perfect alignment of myofibrils in the fiber makes muscle tissue look striped or grainy (like in beef, which is the muscle tissue of a cow or steer); the technical term is *striated*. The light and dark bands of striations repeat along the fiber, and form measurable units called *sarcomeres*. Here's what the three types of muscle tissue are all about:

- ✔ **Cardiac muscle tissue** is found in the heart. Muscle fibers in the heart have just one nucleus; therefore, they're referred to as being *uninucleated*. Cardiac muscle fibers are striated, cylindrical in shape, and branched, like a tree. Large fibers branch off into smaller fibers, which branch off into still smaller fibers. A contraction of the heart muscle must spread quickly throughout the heart, so the cardiac muscle fibers are interlocked. Between contractions, the fibers in the heart muscle relax completely so that the heart muscle does not get fatigued. Because, even when you are fatigued, your heart still needs to beat continually. So, most importantly, contraction of the cardiac muscle is *involuntary*, which means that it occurs without the need for conscious control. Nervous impulses generated from the brain are not necessary to keep the heart muscle contracting. That's just one less thing for your brain to focus on.

- ✔ **Smooth muscle tissue** is found in the walls of hollow internal organs, such as your stomach, bladder, intestines, and lungs. Fibers in this type of muscle tissue are uninucleated and shaped like spindles. They're arranged in parallel lines and form sheets of muscle tissue. Contraction of smooth muscle does not require conscious control; like cardiac muscle tissue, it is involuntary. The contraction in smooth muscle tissue also occurs slowly and stays contracted longer than skeletal muscle tissue. It does not fatigue easily.

- ✔ **Skeletal muscle tissue** is the kind of muscle found in your arms, legs, and torso. These muscle fibers are multinucleated (have many nuclei), and they're striated and cylindrical. Skeletal muscle fibers run the length of the entire muscle, so they can be pretty long, like the hamstrings in the back of your thigh. However, unlike cardiac or smooth muscle tissues, the nervous system controls skeletal muscle, and is therefore under conscious control (it's voluntary rather than involuntary). Although some movements occur quickly — like pulling your hand away from a candle flame if you get burned or throwing your hands in front of your face if something is thrown at you — the nervous system is still receiving input and sending output to make those arm muscles move. (See Chapter 5 for details — take a look at Figure 5-1 to see how skeletal muscles contract.)

# *Getting nervous about nervous tissue?*

Don't be. Just look at Figure 7-3 in Chapter 7 (sorry to make you flip through so many pages) as you read this section, and you'll get the *impulse* to find out more. Nervous tissue forms the nervous system, which is responsible for coordinating the activities and movements of your body through its network of nerves. Parts of the nervous system include the brain, spinal cord, and nerves that branch off of those two key parts. Nervous tissue and nerves are made up of nerve cells, called *neurons.* Neurons are a special type of cell because the nervous system performs some unique functions. Neurons receive and send electrical signals (impulses), they respond to a variety of stimuli (heat, cold, pain, touch, and so on), and they control many activities of the body (such as when hormones should be released). Even for involuntary activities, such as smooth or cardiac muscle contraction, neurons perform the tasks, but impulses from the brain are not necessary to make them do so.

Neurons have cells associated with them that provide support — *neuroglial cells* protect and provide nutrients to the neurons. The neuron itself is made up of a *nerve cell body,* which contains a nucleus and organelles. *Dendrites* branch off the nerve cell body, and they act like little antennae in that they *receive* signals from other cells. The *axon,* which is a long, thin fiber, lies at the other end of the nerve cell body. The axon has branches at its end that *send* signals. How nerves work is explained in Chapter 7.

To distinguish which end of a neuron does what, remember that the axon "acts on" a signal by sending an impulse. A dendrite doesn't send; it receives. If you can remember that "dend" as in dendrite stands for "<u>d</u>oesn't <u>send</u>," then you'll know that dendrites receive and axons send.

# Part II
# Anatomy from Head to Toe

The 5th Wave          By Rich Tennant

"NOW THAT I'VE LIGHTENED UP THE ROOM,..."

## In this part . . .

The three chapters in this part cover the components of the body that form the framework — the chassis, if you will — of your great machine. You find out about the different types of bones, muscles, and skin, and how each of those components receives nutrients and oxygen. You uncover the ways that your bones, muscle, and skin replace worn out cells or repair damaged ones. Also in this part, you begin to look at the pathophysiology of different parts of the body. In these sections, you begin to see what can go wrong in particular areas of the body and how disruptions (such as from illness or a genetic problem) and imbalance among the body's systems can cause disease.

# Chapter 4

# Boning Up on the Skeletal System

• • • • • • • • • • • • • • • • • • • • • • • • • • • • • • • • • • • • • • • • • • •

## In This Chapter

▶ Using those skeletons in your closet

▶ Girdling your innards without an elastic undergarment

▶ Seeing what bones are where

▶ Finding out what your bones do (and how they do it)

▶ Curving the spine in painful ways: lordosis, kyphosis, scoliosis

• • • • • • • • • • • • • • • • • • • • • • • • • • • • • • • • • • • • • • • • • • •

*I*f you have any skeletons in your closet, now is the time to pull them out. I'm not talking about your deep, dark secrets. Seriously! Actually looking at a model skeleton is the best way of figuring out what's connected to what and how and what goes on where.

An adult body contains 206 bones, but the skeletal system contains more than just bones. The skeletal system is made up of bones and joints — as well as the cartilage and ligaments that connect bones and joints together. (See Chapter 3 for information on bone tissue as a type of connective tissue.) The skeleton has two major divisions: *axial* and *appendicular*. The *axial skeleton* consists of the bones that lie along the center axis of the body: the skull, hyoid bone, vertebral column, and rib cage. The *appendicular skeleton* consists of appendages that attach to the axial skeleton: the pectoral girdle, pelvic girdle, and limbs.

## Bony Background

Before I get into what bones are found where, I want to give you some background information about bones in general. By the way, bones in generals are the same as bones in privates. I'm only kidding!

## Reporting for duty: The jobs of your bones

Your skeletal system serves a purpose. Without your parts being connected, you would be a sac of bones and tissues rolling around — not very attractive or practical. Your skeleton provides you with the following benefits:

- ✔ Protection of your internal organs and soft tissues, such as your brain, heart, and lungs
- ✔ Support of your body and organs against the pull of gravity
- ✔ Production of blood cells, which are important for carrying oxygen, fighting off invading germs, and shuttling nutrients and wastes throughout the body
- ✔ Storage for necessary mineral salts, such as calcium
- ✔ Something for your muscles to hold on to so that you can move around

## Classifying the bones in your skeleton

Bones come in different shapes and sizes. Appropriately, many bone type names match what they look like, such as flat bones, long bones, short bones, and irregular bones. Check out Table 4-1 for the differences among the four types of bones.

| Table 4-1 | Characteristics of Bone Types | |
|---|---|---|
| *Bone Type* | *Example Location in the Body* | *Characteristics* |
| Flat | Skull, shoulder blades, ribs, sternum, pelvic bones | Like plates of armor, flat bones protect soft tissues of the brain and organs in the thorax. |
| Long | Arms and legs | Like steel beams, these weight-bearing bones provide structural support. |
| Short | Wrists (carpal bones) and ankles (tarsal bones) | Short bones look like blocks and allow a wider range of movement than larger bones. |
| Irregular | Vertebral column, kneecaps | Irregular bones have a variety of shapes and usually have projections that muscles, tendons, and ligaments can attach to. |

# Describing the structure of a bone

Long bones (see Figure 4-1), such as the femur in your thigh, are usually used as the example when describing bone anatomy because they contain many features, such as those in the list that follows, that are found in a variety of bone shapes and sizes.

- **Compact bone:** This dense, hard layer of bone contains bone cells called *osteocytes* that reside in little chambers called *lacunae*. Lacunae contain protein fibers and mineral deposits (such as the calcium about which you hear so many reminders), and they're arranged in groups of concentric circles called a *Haversian system*. When the little chambers (lacunae) are arranged in a circular pattern, they connect and form little canals called *canaliculi*. The canaliculi connect the lacunae together and link the lacunae to the central *Haversian canal*, which serves as a conduit for the blood vessels from the periosteum to pass through the bone tissue. (Think of electrical wires in your basement being surrounded by metal tubing to protect them. The metal tubing is a conduit.) *Volkmann's canals* connect each Haversian system. The Volkmann's canals run almost perpendicularly to the Haversian canals, and they also serve as wider-diameter conduits through which blood vessels can pass. The Haversian systems ensure that even dense, hard bone has access to the blood vessels of the circulatory system.

- **Diaphysis:** This middle portion or shaft of the bone contains a hollow portion surrounded by compact bone called the *medullary cavity*, inside which is the *marrow*. In adults, the medullary cavity contains *yellow marrow*, which is made mostly of fat (think butter). *Red marrow* produces blood cells in both adults and infants, but infants have more red marrow than do adults. In infants, the medullary cavities of most long bones are filled with red marrow; in adults, red marrow is limited to the spongy bone in the ends of long bones as well as the skull, ribs, vertebrae, and breastbone (sternum). In adults, a process called *hematopoiesis* forms red blood cells in spongy bone.

- **Epiphysis:** Each end of a bone has this area that's involved in bone growth (see the next paragraph). The epiphyis is made of *spongy bone* covered with a layer of *compact bone* (the dense, hard bone) that's surrounded by *articular cartilage* that covers the area where another bone connects (articulates) to form a joint.

- **Periosteum:** This tough, fibrous, connective tissue covers the bones and runs over a bone directly to the ligaments and tendons that attach to the bone. Because the periosteum, ligaments, and tendons connect directly with no clear separation, the periosteum is said to be *continuous* with the ligaments and tendons. The periosteum is the tissue that provides the bone with blood. Blood vessels pass from the periosteum into the bone to provide nutrients and oxygen as well as to remove wastes and transport newly created blood cells. (See Chapter 9 for more on the circulatory system.)

✔ **Spongy bone:** Red blood cells are created in spongy bone, which is strong but light, reducing the overall weight of a bone. Spongy bone contains osteocytes and many trabeculae. Platelike structures that follow stress lines in the bone, trabeculae act like braces, providing support to the bones.

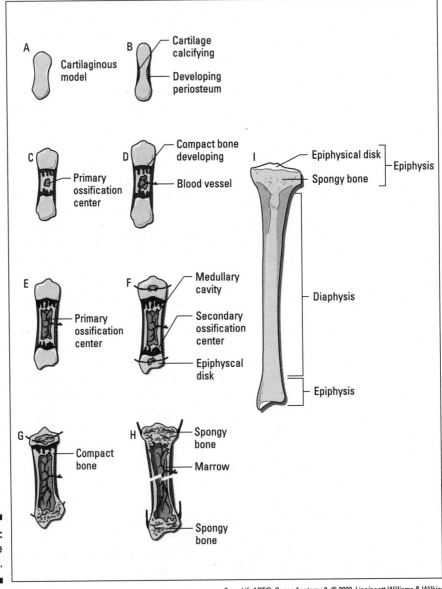

**Figure 4-1:**
Long bone
structure.

Now that you've taken a look at the parts of a bone, keep reading to find out what parts of the bone are involved in making people grow taller.

### My, how you've grown

When you were a wee fetus, snuggled in your mother's womb, you struck some pretty limber poses. With little room to stretch, except for the occasional kick against your mom's bladder, you were curled up tightly for all nine months (unless you couldn't wait and decided to leave earlier). Can you imagine how difficult the fetal position would be if fetal bones were solid? Luckily, as a fetus, you were soft.

When bones develop in a fetus, they're first formed from cartilage. The softer cartilage allows the fetus to bend into the poses that would make a yoga instructor beam with pride. Fetal cartilage takes the shape of the bone it will become, so it serves as a template. Mineral salts, such as calcium, are deposited onto the template, and the cartilage becomes calcified in the *endochondral ossification* process (ossification means "formation of bone").

Remember the medullary cavity? Ossification starts in the middle of the bone. (In Latin, *medullae* means innermost or middle.) Cells that form bone — called *osteoblasts* — continue to calcify the original cartilage cells. As a child ages, the bone cells (osteocytes) replace the osteoblasts that convert the cartilage to bone. Eventually, the only areas of growth are in the epiphyses at the ends of certain bones.

Within each epiphysis, the *epiphyseal disk* contains cartilage. As the cartilage cells divide, the bone lengthens. When the cartilage cells in the epiphyseal disks no longer divide, the bone stops growing in length and adult height is reached.

### Casting a glance at repairing fractures

Fractures are breaks in bones. The body goes through four different steps to repair fractures, but first, the types of fractures that can occur:

- **Comminuted fractures:** The bone is broken into several pieces.

- **Complete fractures:** The bone is broken into two pieces.

- **Compound fractures:** The broken bone acts as a knife and cuts through the skin from the inside out.

- **Greenstick fractures:** The break is incomplete and looks frayed. (Imagine bending a twig that you just tore off a tree; when the stick is green and fresh, it doesn't snap into two pieces.)

- **Impacted fractures:** When a bone breaks into two pieces, the ends are jammed into each other.

✔ **Partial fractures:** The bone is broken along its length but isn't split into two pieces.

✔ **Simple fractures:** The broken bone doesn't poke through the skin.

If you're unlucky enough to have a fracture, as my 6-year-old niece did when she fell off the monkey bars at school, you'll go through a period of repair. Your body does the work all by itself; a cast, rod, and/or pin just keeps the bone still enough for your body to do the job. The following list offers you a general idea of what happens:

1. **A hematoma (blood clot) forms** in the space created by the fracturing of the bone. Blood leaks out of the blood vessels that break as the bone breaks; clotting keeps blood from continuing to flow out of the wounded vessels. The area around the break and hematoma becomes swollen and inflamed because cells from your immune system respond to help prevent infection. (See Chapter 13 for more on the immune system.)

2. **Tissue repair begins** with fibrocartilage filling the space where the break occurred. Fibers of the protein collagen connect the pieces of bone.

3. **A bony callus forms** to join the broken pieces together. This happens when the bone-forming cells — the osteoblasts — create spongy bone complete with trabeculae (the framework).

4. **Remodeling occurs.** The bones don't tear down plaster and put up sheetrock, rewire, and put in new plumbing, but they do put in a kind of new molding. The osteoblasts that create new bone cells lay down a framework of new compact bone cells on the edges of the break. As the new cells are laid down, bone-destroying cells — called *osteoclasts* — reabsorb the spongy bone, creating a new medullary cavity.

Because your body is continually replacing bone cells, you need calcium throughout your entire life, not just when you're a kid. In fact, adults actually seem to need more calcium than do children!

# You're always tearing down and rebuilding

Even when your bones are perfectly intact, your body just can't leave things alone. Change is constant in living things, and those fidgety osteoclasts and osteoblasts need something to do. Just like any other cell in the body, the old, worn-out ones are removed from service and replaced with young, fresh upstarts. The osteoclasts absorb the tired bone cells, and the cellular material is put into the bloodstream. What isn't needed makes its way to the excretory system and is removed as waste; the calcium, however, is recycled through the bloodstream and used by osteoblasts, which create new bone cells. (Personally, I think the term osteo-*blasts* should have been given to the cells that *destroy* old cells — it just makes more sense and would be a great way to remember which cell does what.)

# Centering on Your Axis: The Bones of the Axial Skeleton

The axial skeleton consists of the bones that lie along the midline (center) of your body, such as your vertebral column (backbone). An easy way to remember what bones make up the axial skeleton is to think of the vertebral column running down the middle of your body and then the bones that are directly attached to it — the thoracic cage (rib cage) and the skull. The hyoid bone (see "The bone that floats" sidebar in this chapter), although it's not attached to another bone, is in line with the skull and vertebral column, so it's considered part of the axial skeleton.

## Keeping your head up: The skull

If you think that your skull is one big piece of bone, like a cap, that fits over your brain, you're wrong. The skull not only includes the cranium, which is formed from several bones, but also the bones of the face.

A human skull (see Figure 4-2) is made up of the *cranium* and the *facial bones.* The facial bones surround spaces called the *sinuses,* which do serve a purpose other than being a common site of upper respiratory infections.

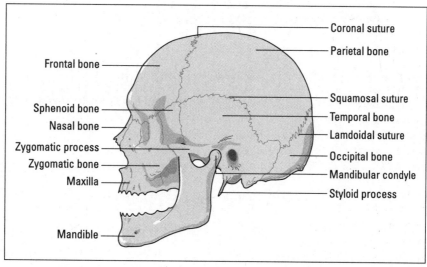

**Figure 4-2:** The human skull: Cranium and facial bones.

*From LifeART®, Super Anatomy 1, © 2002, Lippincott Williams & Wilkins*

### Cracking the cranium

The eight bones of your cranium protect your brain and have immovable joints between them called *sutures*. If you know that a suture is the material used to stitch together an incision or wound, then you can pick out the sutures on a skull. They look like stitches. The bones of the cranium that are joined together by sutures include the following:

- **Ethmoid bone:** Contains several sections, called plates, most of which form the nasal cavity. One of these plates is the *cribriform plate,* and it contains little holes through which nerves from olfactory (sense of smell) receptors pass.

- **Frontal bone:** Gives shape to the forehead, the eye sockets, and part of the nose.

- **Occipital bone:** Forms the back of the skull and the base of the cranium. The *foramen magnum,* an opening in the occipital bone, allows the spinal cord to pass into the skull and join the brain.

- **Parietal bones:** Two bones that form the roof and sides of the cranium.

- **Temporal bones:** Form the sides of the cranium near the temples. The temporal bone on each side of your head contains the following structures:

  - **External auditory meatus:** The opening to your ear canal

  - **Mandibular fossa:** Articulates with the mandible (the lower jaw)

  - **Mastoid process:** Provides a place for neck muscles to join your head

  - **Styloid process:** Serves as an attachment site for muscles of the tongue and voice box (larynx)

- **Sphenoid bone:** Shaped like a butterfly or a saddle (depending on how you look at it), the sphenoid forms the floor of the cranium and the sides of the eye sockets (orbits). A central, sunken portion of the sphenoid bone called the *sella turcica* shelters the pituitary gland, which is very important in controlling major functions of the body. (See Chapter 8 for more on the pituitary gland.)

## What is a cleft palate?

The palate is the roof of your mouth. Part of the palate is soft (way in the back), and part is hard (directly behind your top teeth). Appropriately, the palatine bones form part of your hard palate. A cleft palate occurs when the palatine bones don't fuse during fetal development. With the palatine bones unconnected, an opening exists between the roof of the mouth and the nasal cavity.

## Facing the facial bones

Most of your face is really made up of tissue and muscle. Besides the bones of the cranium that wrap around your face, your face has several cavities (holes), such as the eye sockets (orbits) and nasal cavity. The other fairly small bones that form facial structures are

- **Lacrimal bones:** Two tiny bones on the inside walls of the orbits. A groove between the lacrimal bones in the eye sockets and the nose forms the *nasolacrimal canal.* Tears flow across the eyeball and through that canal into your nasal cavity, which explains why your nose "runs" when you cry.

- **Mandible:** The lower jaw and the only moveable bone of the skull.

- **Maxillae:** Two of these bones form the upper jaw.

- **Nasal bones:** Two rectangular-shaped bones that form the bridge of your nose. The lower, moveable portion of your nose is made of cartilage.

- **Palatine bones:** Form the back of the hard palate (roof of your mouth). On the flip side of the hard palate is the floor of the nasal cavity.

- **Vomer bone:** Joins the ethmoid bone to form the nasal septum — that part of your nose that can be deviated by a strong left hook or, in my case, a Superman leap off the couch when I was 3 years old.

- **Zygomatic bones:** Form the cheekbones and sides of the orbits.

## Sniffing out the sinuses

The sinuses, as you may tell from the name, include more than one structure and allow air into the skull, which lightens the weight of the skull, making it easier to hold your head up. Having air in your sinuses also gives resonance to your voice, which means that when you talk, the sound waves reverberate in your sinuses. The reverberation prolongs and intensifies the sound of your voice; when you have a cold and your sinuses are blocked, your voice sounds dull. Right?

Several types of sinuses are named for their location:

- **Mastoid sinuses** drain into the middle ear; an inflammation here is called *mastoiditis.*

- **Maxillary sinuses** are large and flanked by the bones of the upper jaw (the maxilla).

- **Paranasal sinuses — the frontal, sphenoidal, ethmoidal, and maxillary sinuses —** drain into the nose (para means near; nas- means nose), as I'm sure you noticed whenever you had a cold.

# The bone that floats

The hyoid bone, a tiny U-shaped bone that resides just above your voice box (larynx), anchors the tongue and muscles used during swallowing. However, the hyoid bone itself isn't attached to anything. It's the only bone in the body that does not articulate (connect) with another bone. The hyoid bone hangs by ligaments attached to the styloid processes of the temporal bones.

## *Setting you straight on the curved vertebral column*

The vertebral column (see Figure 4-3) runs nearly the full length of your body. It begins within the skull and extends down to the pelvis. As a section of the body, it contains a fair amount of bones — 33 bones in all. The vertebral column is made up of vertebrae, and each vertebrae is a separate bone.

An important purpose of the vertebral column and all its bones is to protect your spinal cord — the lifeline between your body and your brain. Nearly all your nerves are connected — directly or through networked branches — to the spinal cord, which runs directly into the brain through the opening in the skull called the foramen magnum.

If you look at the spine from the side, you notice that it curves five times: outward, inward, outward, inward, and outward. The curvature of the spine helps it absorb shock and pressure much better than if the spine were straight. A curved spine also affords more balance by better distributing the weight of the skull over the pelvic bones, which is needed to walk upright. A curved spine keeps you from being "top heavy." Each curvature spans a region of the spine: cervical, thoracic, lumbar, sacral, and coccygeal. The number of vertebrae in each region and some important vertebral features are given in Table 4-2.

| Table 4-2 | Regions of the Vertebral Column | |
| --- | --- | --- |
| *Region* | *Number of Vertebrae* | *Features* |
| Cervical | 7 | The skull attaches at the top of this region to the vertebrae called the *atlas*. (In Greek mythology, Atlas held the world on his shoulders; so you can remember that your atlas is the vertebrae that holds your head on your shoulders.) |
| Thoracic | 12 | The ribs attach to this region. |

| Region | Number of Vertebrae | Features |
|---|---|---|
| Lumbar | 5 | Commonly referred to as the small of the back, it takes the most stress. |
| Sacral | 5 (fused into one; the sacrum) | The sacrum forms a joint with the hipbones. |
| Coccygeal | 4 (fused into one; the coccyx, also called the tailbone) | The coccyx is the remains of a tail, which was eliminated during evolution. |

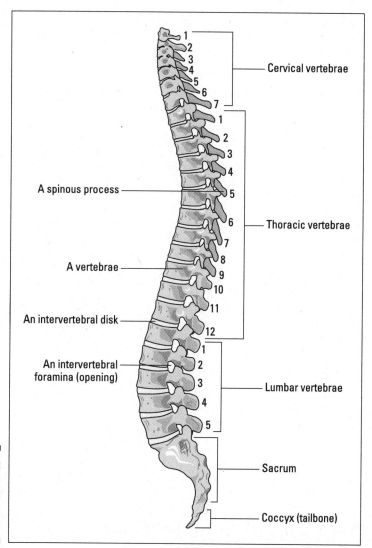

**Figure 4-3:**
The vertebral column, side view.

*From LifeART®, Super Anatomy 1, © 2002, Lippincott Williams & Wilkins*

The vertebral column also provides places for other bones to attach. The skull is attached to the top of the cervical spine. The first cervical vertebra (abbreviated C-1; "C" for cervical, "1" for first) is the *atlas,* which supports the head and allows it to move forward and back. The second cervical vertebra (C-2) is called the axis, and it allows the head to pivot and turn side to side. You can differentiate these two important bones by recalling the Greek story about Atlas, who held the world on his shoulders. Your atlas holds your head on your shoulders.

## *Being caged can be a good thing*

The rib cage (also called the thoracic cage because it's in the thoracic cavity) consists of the thoracic vertebrae, the ribs, and the sternum (see Figure 4-4). The rib cage is essential for protecting your heart and lungs and for providing a place for your shoulder bones to attach.

You have twelve pairs of *bars* in your cage. Some of your ribs are *true,* some are *false,* and some *float.* All ribs are connected to the bones in your back (the thoracic vertebrae). In the front, true ribs connect directly to the sternum (breastbone); false ribs are connected to the *sternum* by *costal cartilage* (*cost-* means rib). The last two pairs of ribs are called floating ribs because they remain unattached in the front. The floating ribs give protection to abdominal organs, such as your kidneys, without hampering the space in your abdomen for the intestines.

The sternum has three parts: the *manubrium,* the *body,* and the *xiphoid* (pronounced *zi*-foid) process. The notch that you can feel at the top center of your chest, in line with your collar bones (the *clavicles*), is the top of the manubrium. The middle part of the sternum is the body, and the lower part of the sternum is the xiphoid process. The xiphoid process is a landmark for cardiopulmonary resuscitation (CPR); if you lay three fingers horizontally above the xiphoid process, that is the spot where CPR should be initiated.

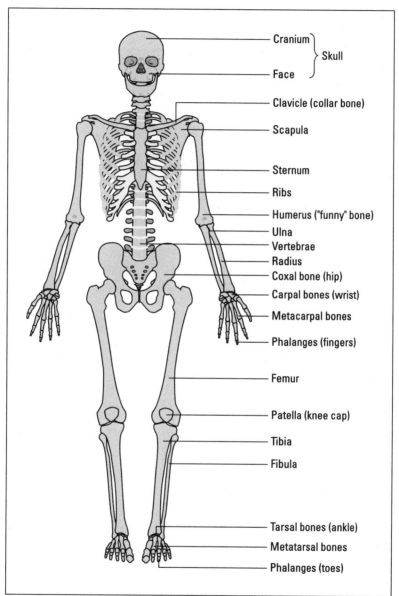

Cranium ⎫
          ⎬ Skull
Face     ⎭

Clavicle (collar bone)

Scapula

Sternum

Ribs

Humerus ("funny" bone)

Ulna

Vertebrae

Radius

Coxal bone (hip)

Carpal bones (wrist)

Metacarpal bones

Phalanges (fingers)

Femur

Patella (knee cap)

Tibia

Fibula

Tarsal bones (ankle)

Metatarsal bones

Phalanges (toes)

**Figure 4-4:**
Front view
of the
skeleton
showing the
rib cage,
clavicles,
upper limbs,
lower limbs,
and pelvis.

*From LifeART®, Super Anatomy 1, © 2002, Lippincott Williams & Wilkins*

# Making a Connection: The Appendicular Skeleton

The appendicular skeleton is made up of the appendages and associated bones that join the axial skeleton (see above). These bones include the two "girdles" — pectoral and pelvic — as well as the bones of the arms and legs.

## Wearing girdles: Everybody has two

Your body contains two girdles (the pectoral girdle and the pelvic girdle), and like the girdles your grandmother may have worn, these girdles hold in body parts. The word *girdle* means to encircle. The girdles women wear encircle their waists; the girdles in your body encircle the vertebral column. The two girdles in the body as well as the limbs that are attached to the girdles form the appendicular skeleton.

### Girdling shoulders

This section of your body encircles the top of the vertebral column and provides a place of attachment for your upper limbs, which are better known as your arms. The pectoral girdle (refer to Figure 4-4) consists of the two clavicles that make up your collar bones and the two triangle-shaped scapulae (shoulder blades), also known as chicken wings. The scapulae provide a broad surface to which arm and chest muscles attach.

The collar bones are the only part of the pectoral girdle that are attached to the axial skeleton. (The collar bones meet the top of the sternum — the manubrium.) Because the pectoral girdle is only weakly attached to the axial skeleton, structures of the pectoral girdle (such as the shoulders) have a wide range of motion but are prone to dislocation.

## Why weight-bearing exercise is good for you

You probably know that aerobic exercise and lifting weights are good for your heart and your muscles, but did you realize it benefits your bones, too? Exercise — especially weight-bearing exercise (such as exercises that use the hips and legs: walking, running, bicycling, weight lifting) — increases the activity of osteoblasts, regardless of your age. Osteoblasts are the bone-building cells that mature into osteocytes. So, if you are forming more bone cells, your bones are getting stronger. Exercise staves off osteoporosis, which is the depletion of osteocytes (bone cells) and the weakening of bones. And exercising the muscles also exercises the bones and joints, which maintains flexibility and strength. There's nothing like a firm foundation.

# Why women have bigger hips than men

Okay. You know it's true. Women aren't built like men. Most men tend to be straight up and down with few curves. Women, on the other hand, are hourglasslike — their hips tend to be wider than men's hips. In a woman, the iliac bones flare wider than they do in a man. And the *true pelvis* of a woman — the ring formed by the pubic bones, ischium, lower part of the ilium, and the sacrum — is wider and more rounded. The true pelvis of a man is shaped somewhat like a funnel.

These differences in anatomy have a physiological purpose: Women have babies, and when those babies are ready to be born, they need to pass through a woman's true pelvis without getting stuck. Other differences also relate to giving birth: The sacrum in women is wider and tilted back more so than in men; the coccyx of women moves easier than it does in men. These two features allow a little more "give" when a baby is passing through the pelvis. When a woman is pregnant, she creates a hormone called *relaxin* that allows the ligaments connecting the pelvic bones to relax a little, thus the bones can spread farther apart and flex a bit during delivery.

All these female characteristics are wonderful for giving birth, but as a woman who has done so several times, my experience leads me to believe that the bones don't return to their original positions as readily. The hips tend to stay a bit wider after giving birth. Maybe the bones staying loose and spreading out is a physiological "reward," making delivery of a subsequent infant even easier. Too bad our bodies don't know when that last infant has been delivered, so our hips can go back to their pre-pregnant sizes!

## Girdling hips

Elvis moved this part of the body. The pelvic girdle (see pelvis in Figure 4-4) is formed by the hipbones (called coxal bones), the sacrum, and the coccyx (tailbone). The hipbones bear the weight of the body, so they must be strong.

The hipbones (coxal bones) are formed by the *ilium*, the *ischium*, and the *pubis*. The ilium are what you probably think of as your hipbones; they're the large, flared parts that you can feel on your sides. The part that you can feel is the tip of the hip is the *iliac crest*. In your lower back, the ilium connects with the vertebral column at the sacrum; the joint that's formed is appropriately called the *sacroiliac joint* — a point of woe for many people with lower back pain. The problem could be at the sacroiliac joint itself, such as arthritis or misaligned bones, but another back pain culprit resides in this area as well. The *greater sciatic notch* is an indentation that allows the blood vessels and the large sciatic nerve to pass down into the leg. When the sciatic nerve becomes compressed, the painful condition called *sciatica* can occur.

The *ischium* is the back part of your hip. You have an ischium on each side, within each buttock. You are most likely sitting on your *ischial tuberosity* right now. These parts of your hips are also called the "sitz bones" because they allow you to sit. The ischial tuberosity points outward, and the *ischial spine* — which is around the area where the ilium and ischium join — is directed inward into the pelvis. The distance between a woman's ischial spines is key to her success in delivering an infant vaginally (see Chapters 14 and 15); the opening between the ischial spines must be large enough for a newborn's head to pass through.

# Going out on a limb: Arms and legs

Your arms and legs are "limbs" or "appendages." The word *append* means to attach something to a larger body. One example is an appendix, which is an attachment at the end of a book. As attachments to your larger body, your appendages and the ring of bones called girdles with which they connect make up your appendicular skeleton.

### Giving you a hand with hands (and arms and elbows)

Your upper limb or arm is connected to your pectoral girdle (refer to Figure 4-4). The bones of your upper limb include the *humerus* of the upper arm, the *radius* and *ulna* of the forearm, and the hand, which is made up of *the carpal* and *metacarpal bones* as well as the *phalanges*.

Your humerus, or upper arm bone, connects to your shoulder blade (scapula). The scapula has a feature called the *glenoid cavity,* a depression into which the *head* of the humerus fits. Muscles that move the arm and shoulder attach to the *greater* and *lesser tubercles,* two points near the head of the humerus. The greater tubercle is larger than the lesser tubercle. Between the greater and lesser tubercles is the *intertubercular groove,* which holds the tendon of the biceps muscle to the humerus bone. The humerus also attaches to the deltoid muscle of the shoulder at a point called the *deltoid tuberosity.* The muscle attached to the deltoid tuberosity allows you to raise and lower your arm.

The bones of the forearm attach at the elbow end of your humerus in four different spots:

- **Capitulum:** Condyle (knob) that allows the forearm's radius to articulate with the humerus.
- **Trochlea:** Condyle on the humerus that lies next to the capitulum and allows the *trochlear notch* of the forearm's ulna to articulate with the humerus.
- **Coronoid fossa:** Depression in the humerus that accepts a projection of the ulna bone (called the *coronoid process*) when the elbow is bent.
- **Olecranon fossa:** Depression in the humerus that accepts a projection of the ulna (called the *olecranon* process) when the arm is extended. Fitting, isn't it?

So how do you keep the radius and ulna straight? Well, keep your arm straight with your palm facing forward, and I'll tell you. The radius is the bone on the thumb side of your arm. When you turn your arm so that your palm is facing backward, the radius crosses over the ulna so that the radius can stay on the thumb side of your arm. The radius is shorter but thicker than the ulna. The head of the radius looks like the head of a nail. The ulna is long and thin, and its head is at the opposite end of the bone compared with the head of the radius.

Both the radius and ulna connect with the bones of the hand at the wrist. The wrist contains eight small, irregularly shaped bones called the *carpal bones.* The ligaments binding the carpal bones are very tight, but the numerous bones allow the wrist to flex easily. The eight carpal bones are the *pisiform, triangular, lunate, scaphoid, trapezium, trapezoid, capitate,* and *hamate.* The palm of your hand contains five bones called the *metacarpals.* When you make a fist, you can see the ends of the metacarpals as your knuckles. Your fingers are made up of bones called the *phalanges;* each finger has three phalanges (phalanx is singular): the *proximal phalanx,* which joins your knuckle, the *middle phalanx,* and the *distal phalanx,* which is the bone in your fingertip. The thumb has two phalanges, so sometimes it's not considered a true finger. So you may have eight fingers and two thumbs or ten fingers depending on how you look at it.

### Getting a leg up on your lower limbs

Your lower limb consists of the *femur* (thigh bone), the *tibia* and *fibula* of the lower leg, and the bones of the foot: *tarsals, metatarsals,* and *phalanges* (refer to Figure 4-4).

The term *phalanges* refers to the finger bones *and* the toe bones.

The femur is the strongest bone in the body; it's also the longest. The head of the femur fits into a hollowed out area of the hip bone called the *acetabulum.* In women, the acetabula (plural) are smaller but spread farther apart than in men. This anatomic feature allows women to have a greater range of movement of the thighs than men. The *greater and lesser trochanters* of the femur are surfaces to which the muscles of the legs and buttocks attach. Trochanters are large processes found only on the femur. The *linea aspera* is a curve along the back of the femur to which several muscles attach.

The femur forms the knee along with bones of the lower leg. The *patella* (commonly known as the kneecap) articulates with the bottom of the femur. The femur also has knobs *(lateral and medial condyles)* that articulate with the top of the tibia. The ligaments of the patella attach to the tibial tuberosity. The bottom, inner end of the tibia has a bulge called the *medial malleolus,* which forms part of the inner ankle.

## Strong feet, firm foundation

The arches of your feet help to absorb the shock created by the pounding of your feet when walking or running, and they help to distribute your weight evenly to the bones that bear the brunt of it: the calcaneus (heel) and talus (ankle) in each foot. The heel, ankle, and metatarsals bear a significant amount of your body weight, and are thus called "weight-bearing bones." If the ligaments and tendons that form those arches along with the tarsals and metatarsals weaken, the arches can collapse, resulting in flat feet. A person with flat feet is more prone to damage to the bones of the feet (such as the metatarsals) because of increased pressure placed on those bones. Examples of damage include bunions, which is a painful displacement of the first metatarsal (the big toe), and heel spurs, which are bony outgrowths on the calcaneus that cause pain when walking. Flat feet also can contribute to knee pain, hip pain, and lower back pain.

The tibia, also called the shinbone, is much thicker than the fibula and lies on the inside (medial) portion of the lower leg. Although the fibula is thinner, it's about the same length as the tibia. The bottom, outside end of the fibula is the *lateral malleolus,* and it is the bulge of the outside of your ankle.

Your foot is designed in much the same way as your hand. The ankle, which is akin to the wrist, consists of seven tarsal bones. Altogether, the bones of the ankle are called the *tarsus,* but only one of those seven bones is part of a joint with a great range of motion — the *talus.* The talus bone joins to the tibia and fibula and allows your ankle to rotate. The largest tarsal bone is the *calcaneus,* which is the heel bone. The calcaneus and talus help to support your body weight.

The instep of your foot is akin to the palm of your hand, and just as the hand has carpals and metacarpals, the foot has tarsals and metatarsals. The ends of the metatarsals on the bottom of your foot form the ball of your foot. As such, the metatarsals also help to support your body weight. Together, the tarsals and metatarsals held together by ligaments and tendons form the arches of your feet. Your toes are also called phalanges, just like your fingers. And, just as your thumbs have only two phalanges, your big toes have only two phalanges. But, the rest of your toes have three: proximal, middle, and distal.

# Articulating Joints

Another word for joints is "articulations" from the Latin word *articularis,* meaning "jointed." There are many types of joints, and they tend to be classified by the amount of movement they permit. This section tells you about the different joint structures and about the movements they allow.

# *Seeing which joints jump*

You probably think of joints as the hingelike type in your knees, but a joint is simply a connection between two bones. Some joints move freely, some move a little, and some never move.

### *Joining immovable bones*

*Synarthroses* are joints that do not move. Examples of joints in the body that never move are the synarthroses in the skull. The sutures between the different bones in the cranium do not move. A thin layer of connective tissue joins them together. The sutures in the cranium include

- ✔ **Coronal suture:** Joins the parietal bones and the frontal bone
- ✔ **Lambdoidal suture:** Joins the parietal bones and the occipital bone
- ✔ **Sagittal suture:** Between the parietal bones
- ✔ **Squamosal sutures:** Between the parietal and temporal bones

### *Joining bones that move a little*

*Amphiarthroses* are slightly moveable joints connected by fibrous cartilage (fibrocartilage) or hyaline cartilage (see Chapter 3). Examples include the vertebrae of the spinal column. The intervertebal disks join each vertebrae and allow slight movement of the vertebrae.

### *Joining bones that move freely*

*Diarthroses* are the type that probably come to mind when you think of joints that are freely moveable (Table 4-3). Diarthroses are also synovial joints because a cavity between the two connecting bones is lined with a synovial membrane and filled with *synovial fluid,* which helps to lubricate and cushion the joint.

Diarthroses are joined together by *ligaments,* which are made of fibrous connective tissue. *Tendons* are fibrous connective tissue that join muscles to bones. Tendons also help to stabilize joints, but they do not form joints. *Bursae* are fluid-filled sacs that help to reduce the friction between the tendons and ligaments and between the tendons and bones. The knee contains 13 bursae; inflammation in these sacs is called *bursitis.* What is commonly called tennis elbow is bursitis in the elbow.

| Table 4-3 | Types of Diarthroses (Synovial Joints) | | |
|---|---|---|---|
| *Type of Joint* | *Description* | *Movement* | *Example* |
| Ball-and-socket joint | The ball-shaped head of one bone fits into a depression(socket) in another bone | Circular movements; joints can move in all planes,and rotation is possible | Shoulder, hip |
| Condyloid joint | Oval-shaped condyle of one bone fits into oval-shaped cavity of another bone. | Can move in different planes but cannot rotate | Knuckles (joints between meta-carpals and phalanges) |
| Gliding joint | Flat or slightly curved surfaces join | Sliding or twisting in different planes | Joints between carpal bones (wrist) and between tarsal bones (ankle) |
| Hinge joint | Convex surface joins with concave surface | Up and down motion in one plane | Elbow, knee |
| Pivot joint | Cylinder-shaped projection on one bone is surrounded by a ring of another bone and ligament | Rotation is only movement possible | Joint between radius and ulna at elbow and joint between atlas and axis at top of vertebral column |
| Saddle joint | Each bone is saddle shaped and fits into the saddle-shaped region of the opposite bone | Many movements are possible | Joint between carpal and meta-carpal bones of the thumb |

# Knowing what your joints can do

You know that certain types of joints can perform certain kinds of movements. The following list is a quick overview of those special movements. The two basic types of movements are angular and circular.

*Angular movements* make the angle formed by two bones larger or smaller. Examples of these include:

- ✔ **Abduction** moves a body part to the side, away from the middle of the body. When you make a snow angel, and you move your arms and legs out and up, that's abduction.

- ✔ **Adduction** moves a body part from the side toward the middle of the body. When you're in snow angel position, and you move your arms and legs back down, that's adduction.

- ✔ **Extension** makes the angle larger. Hyperextension occurs when the body part moves beyond a straight line (180 degrees).

- ✔ **Flexion** decreases the joint angle. When you flex your arm, you move your forearm to your upper arm.

*Circular movements* only occur at ball-and-socket joints like in the hip or shoulder. Examples include the following:

- ✔ **Circumduction** is the movement of a body part in circles.

- ✔ **Depression** is the downward movement of a body part.

- ✔ **Elevation** is the upward movement of a body part, such as shrugging your shoulders.

- ✔ **Eversion** only happens in the feet when the foot is turned so the sole is outward.

- ✔ **Inversion** also only happens in the feet when the foot is turned so that the sole is facing inward.

- ✔ **Rotation** is the movement of a body part around its own axis, such as shaking your head to answer, "No."

- ✔ **Supination** and **pronation** refer to the arm and stem from the terms supine and prone. Supination is the rotation of the lower arm to make the palm face upward or forward. Pronation is the rotation of the lower arm to make the palm face downward or backward.

# Pathophysiology of the Skeletal System

Although bones are incredibly strong, they're prone to injuries, the effects of aging, and disease, just like any other body part. This section gives you some information on a few of the most common problems that occur in bones or joints.

# Are you twisted sister?

Abnormal curvatures of the spine can cause plenty of pain and can lead to several problems. When the curve of the lumbar spine is exaggerated, the abnormal condition is *lordosis,* more commonly known as *swayback.* The lumbar spine of a pregnant woman becomes exaggerated because the woman needs to balance the pregnant belly on her frame. However, sometimes the curve remains after pregnancy when weakened abdominal muscles fail to support the lumbar spine in its normal position. Developing the habit of holding the abdominal muscles in (rather than letting it all hang out, so to speak) helps to strengthen the center of the body and prevent swayback. Losing the beer belly helps, too.

Older men and women sometimes develop an abnormally curved spine in the thoracic region — a condition called *kyphosis* — what's commonly known as *hunchback.* Osteoporosis or normal degeneration and compression of the vertebrae tends to straighten the cervical and lumbar regions of the spine and push out the thoracic vertebrae, thus causing kyphosis.

You may recall being checked for *scoliosis* during junior high gym class. The reason for that inspection is because scoliosis (twisted spine) first becomes obvious during the late childhood/early teen years — just when people are most self conscious. When you look at the spine from the back, it appears to be straight — the normal curvature is evident when you view the spine from the side. However, in people with scoliosis, the spine curves side to side and looks S-shaped when viewed from the back.

# Out and about with gout?

*Gout* is a metabolic disease, but it affects the joints by making them red, swollen, and painful. Gout is caused by deposits of uric acid (which normally are removed in urine) in the joints. The condition of having too much uric acid in the bloodstream is *hyperuricemia,* which contributes to gout, but what causes hyperuricemia is unknown. The bloodstream can contain an over-abundance of uric acid because a genetic defect causes the body's cells to produce too much uric acid or the kidneys aren't excreting as much as they should. Whatever the cause, the resulting gout is unpleasant.

As the uric acid is deposited into the joint spaces, it crystallizes. Initially, there are no symptoms. But, eventually, the joints become so full of the crystals that the joint becomes painful and swollen. After the first attack, a person with gout may go for months or years before having another attack. If the gout is not treated, the uric acid crystals continue to fill the joints, potentially damaging cartilage, synovial membranes, tendons, and soft tissues such as the muscles adjoining the bone.

Eventually, hard, yellow nodules called *tophi* develop. The tophi usually appear on the big toe or outer ear, but can occur on hands, knees, forearms, or the Achilles tendon on the back of the ankle. In the severe, chronic form, tophi can cause deformities and limited movement of the joint. Complications include kidney stones, nerve damage, and circulatory problems.

Gout is treatable with medications, such as colchicine, corticosteroids, and allopurinol. Lifestyle changes such as weight reduction, drinking plenty of water, and avoidance of alcohol also help a great deal.

## Pouring over osteoporosis

When I was in college, my family had a reunion at my Great-Aunt Lucy's cabin in the Poconos. My great-aunt was never tall, but I was shocked when I saw her that morning; she had shrunk! I hadn't grown much, if at all, since our previous visit, but she had definitely lost inches off her height. Perhaps she had osteoporosis.

Osteoporosis is a disease that results in an imbalance between the breaking down of bone (bone resorption) and the building up of bone (bone formation). Some causes of osteoporosis include

- ✔ Aging
- ✔ Alcoholism
- ✔ Hormonal imbalance
- ✔ Poor diet
- ✔ Prolonged use of steroids
- ✔ Rheumatoid arthritis
- ✔ Sedentary lifestyle

However, osteoporosis occurs most often in postmenopausal women. After menopause, women produce less estrogen, so they lose the protective effect of estrogen on the bones; estrogen keeps the bones from losing calcium. In people with osteoporosis, calcium is pulled from the bones, but it's not recirculated to help build new osteocytes. Because the bone cells are being broken down faster than new bone cells are being created, the supporting structures of the bones — remember the trabeculae in spongy bone? — deteriorate. The deterioration leads to fractures.

As tiny fractures occur in the vertebrae of the spinal column, the vertebrae compress, and the curvature of the spine can change, which can lead to a loss of height. After the curvature of the spine changes, the abdomen can begin to protrude as the body attempts to adjust to a new center of a gravity.

Osteoporosis really can't be cured. After it's diagnosed, further bone loss can be controlled and the pain can be eased. But it's much more effective to try to prevent osteoporosis through consuming a balanced diet with plenty of vitamin D, calcium, and protein, and plenty of weight-bearing exercise to help create new bone cells. (See the sidebar on "Why weight-bearing exercise is good for you" in this chapter.) Prevention should start in early adulthood, especially for women with a family history of osteoporosis.

# Chapter 5

# Bulking You Up on Muscles

*W*ithout muscles, where would you be? In a heap on the floor. Without muscles, your bones don't connect. You can't move. You'd be dead. See, your heart is a muscle, and without the heart to pump blood through your body, you're a goner. Muscles are extremely important to your anatomy, and their physiology runs your physiology. In this chapter, you find out about the different types of muscle tissue, how muscles contract, what muscles are in your body, and what can go wrong in the muscular system.

## Making the Most of Your Muscles

You know those muscle types, the guys at the gym that hover around the free weights donned in tight-fitting, sleeveless T-shirts. Some have gold chains, some have golden tans, and some even have gold teeth. This section is not about them. It's about something much more golden — your ability to move. Being able to flex and ripple may be the goal of some people, but just being able to flex and extend enough to walk is the goal of some others. Muscles allow you to do some important things:

   ✓ **Muscles allow you to stand upright.** Gravity is a strong force. If your muscles don't contract, the force of gravity can keep you down in that heap on the floor. Muscle contraction requires that the force of gravity be opposed. Strength is a measurement of how much gravitational force or weight that your muscles can oppose.

- ✔ **Muscles allow you to move.** The fact that muscles allow you to move as in walking or running may seem obvious, but muscle contraction allows your body to assume different positions, too. Can you imagine being as stiff as the Tin Man from the *Wizard of Oz*? Muscles allow for every little movement right down to the blink of an eye, the dilation of your pupils, and smiling.

- ✔ **Muscles allow you to digest and control waste removal.** The organs in your digestive system are lined with muscles that keep the food moving downward and outward. The contraction of muscles creates *peristalsis,* the rhythmic squeezing of food down through the esophagus, stomach, and intestines. Contracted sphincter muscles hold urine in your bladder and feces in your colon until you are ready to excrete them. When you relax those sphincter muscles, you allow the waste to be released whenever and wherever you are ready.

- ✔ **Muscles affect the rate of blood flow.** Blood vessels, such as arteries and veins, are lined with muscle tissue that enables them to dilate to allow blood to flow faster through the vessel or to contract, which slows down the rate of blood flow. Muscle contraction throughout your body is also responsible for moving blood through your veins back to your heart, and then your heart muscle pumps the blood through the arteries.

- ✔ **Muscles hold your skeleton together.** Ligaments and tendons at the ends of the muscles wrap around joints and hold them together, thus holding together the bones of the skeleton, too.

- ✔ **Muscles help to maintain normal body temperature.** Muscle contraction is a physiologic process and like most physiologic processes involves some chemical reactions. In the chemical reactions that occur to cause muscle to contract, heat is released that is used to maintain body temperature because some heat is continually lost through your skin. When you get cold, your body shivers. Muscle contraction causes shivering in an attempt to generate heat. Think about it. In winter, you stay warm when you are skiing and ice skating outside in the cold air, but you feel cold when you are just sitting still inside your home. If you're cold, get moving!

# Talking About Tissue Types

Three basic muscle tissue types allow you to do all that's crucial to living: *cardiac, smooth,* and *skeletal.*

- ✔ **Cardiac muscle tissue:** Found in the heart, the fibers in cardiac muscle contain one nucleus (so they're *uni*nucleated — *uni*- means "one"); they're striated (striped), and their shape is cylindrical and branched. Cardiac muscle tissue fibers interlock, which allows the contraction to spread quickly throughout the heart. Between contractions, the fibers relax completely so that the heart muscle does not wear out. Luckily, contraction of the heart muscle is completely involuntary, which means it occurs without stimulation by a nerve and happens without you even being conscious of it. You don't have to think about making your heart beat. Contraction of cardiac muscle is described in Chapter 9.

✔ **Smooth muscle tissue:** Existing in the walls of internal organs that are hollow, such as the stomach, bladder, intestines, and lungs, these fibers also are uninucleated, but they're shaped like spindles and arranged in parallel lines. Their shape allows them to form sheets of muscle tissue. Smooth muscle contraction is involuntary. You don't have to think about your digestive system moving food from one organ to the next. When it contracts, it does so slowly, so it can stay contracted longer than a skeletal muscle. It doesn't fatigue easily because smooth muscle contracts slowly. Smooth muscle as part of the digestive system is discussed in Chapter 11.

✔ **Skeletal muscle tissue:** Typically what people picture when they think of a "muscle" is the skeletal muscle tissue comprising your bulging biceps, your planklike pecs, and your hard-as-rock hamstrings. Skeletal muscles are the ones that hold your skeleton together and make it move. The skeletal muscle fibers have many nuclei (they're *multi*nucleated), and they're striated. Their cylindrical shape allows them to run the length of a muscle; therefore, some muscle fibers are long. The remainder of this chapter focuses on skeletal muscles.

Meat is an example of muscle fibers running the length of a muscle. (Meat, after all, is muscle tissue.)

# Getting a Grip on Skeletal Muscle Contractions

The nervous system controls the skeletal muscles. (See Chapter 7 for more on the nervous system.) Sometimes the reactions are involuntary, like when you touch something hot or sharp with your finger, and your nervous system sends a signal to the muscles in your arm to pull your hand away. (See Chapter 7 for more on reflex arcs.) At other times, your skeletal muscles are under conscious (voluntary) control, such as when you decide that you are going to rush the net during a tennis match and hit a nice angle shot to score match point. Muscles are required to make you run in the direction that you want to go and to swing the racquet at the right time. Figure 5-1 shows you how skeletal muscle is connected to the nervous system, as well as how skeletal muscle contracts.

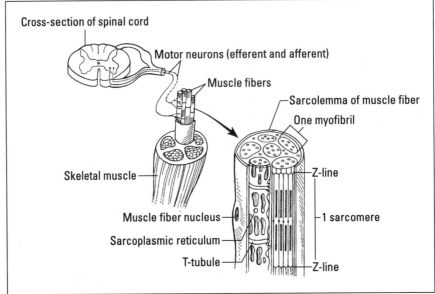

**Figure 5-1:**
Anatomy of
a skeletal
muscle. The
muscle's
connection
to the spinal
cord (left)
and a close-
up diagram
of a muscle
fiber (right).

Muscle contractions require the work of quite a few components. Luckily, in your day-to-day activities, you don't have to worry about each component doing its job — it just does it! I describe those components, what they are and what they do, and then I explain how those components work together to make a contraction happen in the next two sections.

## Meeting the components that cause a contraction

Two basic components cause a muscle contraction: the structure of the muscle and ATP (*adenosine triphosphate* — see Chapter 3), the chemical compound that gives the muscle the boost it needs to make the contraction occur. The following two sections take a look at each of these two basic components.

### Muscling in on muscles

Fibers make up muscles. (Refer to Figure 5-1 as you read this section.) Think of a muscle fiber as a long, thin strand. Within that strand are myofibrils, which are just as long, but even thinner than the muscle fiber. Myofibrils are perfectly aligned, which make the muscle look striped (the technical term is *striated*). Each myofibril contains even smaller *thin filaments* and *thick filaments*.

- ✔ **Thin filaments** of a myofibril contain two strands of *actin,* which is a double helix-shaped protein (similar in shape to DNA). Along actin's double helix are binding sites for molecules of *troponin* and *tropomyosin.*

- ✔ **Thick filaments** have groups of the protein *myosin.* The end of a myosin strand has a bulb-shaped structure. In thick filaments, some myosin strands lie in one direction, and some lie in the opposite direction. You can identify a thick strand because the mixed strands make thick filaments look like they have bulb-shaped structures at both ends.

One unit of striation, called a *sarcomere,* occurs between the light and dark bands, and myofibrils are filled with sarcomere after sarcomere.

The actin and myosin filaments of a myofibril line up like good little soldiers. The actin filaments are attached to the outer edge of a light band, called a Z line. Dark myosin filaments come next, and they're unattached to a Z line. Next, another actin filament. Think of a split-rail fence; the Z lines are the posts (vertical) and the actin filaments are the rails (horizontal) attached to the posts. But just to make it a bit tricky, each actin filament does not connect from Z line to Z line unbroken. A gap, called the *H zone,* exists in the middle of each sarcomere. The myosin filaments are located between each actin filament and appear to "float" because they're unattached to the Z line. This alternating pattern of light and dark bands repeats throughout the length of a myofibril. One *sarcomere* (the unit of contraction) runs from the outer edge of one light band (Z line) to the outer edge of the next light band (Z line). From Z line to Z line is one sarcomere.

The variety of proteins in muscle tissue is what makes meat such a good source of protein in your diet.

### Energizing your muscles: ATP

Cells convert the energy in the food you eat into ATP (adenosine triphosphate), the "currency" of the cell. (For more on cells, see Chapter 3.) Most cells in your body — except brain cells — can't use glucose directly. (See Chapters 2 and 3 for more on glucose.) They must convert glucose to a useable form of energy, which is ATP. Then the cell uses ATP to run the chemical reactions that keep the cell metabolizing.

ATP is also required to induce the chemical reaction for muscle contraction. Muscle fibers contain enough ATP to sustain a contraction for about one second. So you need a steady supply of ATP to keep muscle contractions going. Whether you realize it, you always have some muscles that are contracted; total relaxation is a myth. And don't forget that your heart is a muscle, and it, too, needs ATP.

When your supply of oxygen is used up, you're said to be "oxygen depleted." Without oxygen, your body cannot create ATP using the normal pathway of aerobic cellular respiration (see Chapter 2), but you still need a supply of ATP. So your body switches to a back-up plan that uses lactic acid to create ATP.

Lactic acid is produced during anaerobic (without oxygen) respiration (see Chapter 2), but muscles can't use it directly for contraction. The energy is stored in molecules of ATP plus *creatine* (called *phosphocreatine*), which form during muscle relaxation periods. By "muscle relaxation," I mean the short time between contractions; the muscle is not off on a Caribbean island sunning itself on the beach with a rum-based drink or lying on a massage table in a totally rejuvenating spa. It does sound wonderful, though, doesn't it? Okay. Back to muscle physiology.

Phosphocreatine breaks down quickly to release more ATP when needed. Humans are resourceful, however, and also take every two molecules of *adenosine diphosphate* (ADP; two phosphates each for a total of four) that form during a contraction and make new ATP. The new ATP uses up three of the four phosphate molecules, and the leftover one goes toward making a molecule of the one-phosphate substance *adenosine monophosphate* (AMP). When the amount of AMP rises in the cell, *glycolysis* (see Chapter 2) is stimulated to synthesize more ATP. In short, your body does everything it can to ensure that some ATP is always in your cells.

## Getting in on the action

The *sliding filament theory* describes muscle contraction. I know, what's a sliding filament? Well, glad you asked.

During muscle contraction, ATP binds to one of the bulb-shaped ends of a myosin filament. When it does, the ATP splits into ADP plus a molecule of inorganic phosphate ($P_i$). The ADP and $P_i$ stay attached to the myosin filament.

Remember the troponin and tropomyosin molecules that I explained in the "Muscling in on muscles" section? They do have a purpose. Calcium ions that bind to the troponin molecules causes the tropomyosin to move out of the way so that the binding sites on the actin filament are exposed.

An *ion* is an electrically charged atom or group of atoms. The electrical charge results when the atom loses or gains one or more electrons. The loss of an electron causes an ion to have a positive charge, and the gain of an electron creates an ion with a negative charge. (See Chapter 2 for more on electrons.)

# The last contraction

The time comes for every animal — humans included — to die. The fact that every animal gets cold and stiff tells others when that time has come. Do you know why? The cells no longer make ATP. At the moment of death, the lungs stop filling with oxygen, the heart stops pumping blood through the body, and the brain stops sending signals. The cells — without incoming oxygen, nutrients, or stimulus from the brain — cease performing their metabolic reactions. So ATP can no longer be produced. Without ATP flooding the myofibrils, contractions cannot occur, but neither can the last step of muscle contraction. In order for a myofibril to relax, ATP must hook onto myosin and dissolve the actin–myosin cross-bridges. But when ATP is unavailable to generate a subsequent contraction, the last contraction becomes permanent, and the corpse stiffens. *Rigor mortis*, which means rigidity of death, occurs in every muscle throughout the body. And remember that movement of muscles generates heat; so when the muscles stop their physiologic reactions and warm blood stops flowing through the blood vessels, the corpse gets cold.

After the binding sites on actin are open, myosin binds to the actin. Cross-bridges form to connect actin and myosin. But the myosin has to give up something so that it can "hold onto" the actin. At this point, it releases the ADP and $P_i$ it held onto after the ATP split.

After myosin releases the ADP and $P_i$, the bulb-shaped end on the myosin filament changes. As it changes, the actin filament "slides" toward the middle of the sarcomere, pulling the Z lines at the end of the sarcomere closer together, and causing the H zone to disappear. This action shortens and contracts the muscle fibers.

When another ATP molecule hooks onto the bulb-shaped end of myosin, the cross-bridges that link the actin and myosin dissolve, and the process begins again. Amazingly, it took you hundreds of times longer to read the last page and a half describing what happens when muscles contract than it takes for a muscle fiber to actually contract.

Another reason to get enough calcium: Muscles don't contract properly without calcium ions. So drink your milk, eat your spinach, and have some yogurt all for a good cause — your own strength.

## Getting to the non-action

When you move, your muscles contract by shortening. (Refer to the "Energizing your muscles: ATP" section earlier in this chapter.) These types of contractions in which a movement occurs is an *isotonic contraction.* However, sometimes you contract your muscles but don't move a body part. Have you hung on a monkey bar lately? If so, that hanging action requires that your biceps muscle tightens up, but your arm doesn't move. That type of contraction in which no movement occurs is an *isometric contraction.* If you tighten up the gluteus maximus muscles in your buttocks as you sit and read this book, you are performing an isometric contraction. But if you move your arm to pick up a cup of coffee or a pen, your biceps muscle performs an isotonic contraction.

## Toning up your knowledge of muscle tone

No matter what you're doing, some of your muscle fibers are contracted. This constant state of contraction is *muscle tone.* Without muscle tone, you'd be a heap on the floor. Seriously. If your muscles were to totally relax all at the same time, you would collapse. So to prevent that embarrassing situation, some of your muscle fibers remain contracted. Muscle tone allows you to maintain correct posture — that is, standing up erect with shoulders back, head up, abdomen in, and knees unlocked.

To keep muscle tone, your muscles rely on *muscle spindles,* which are fibers in your muscles that act like sense receptors. The muscle spindles are specialized muscle fibers that are wrapped with nerve fibers. The central nervous system (brain and spinal cord) stays in contact with the muscles through the muscle spindles. The muscle spindles send messages about your body position through the spinal cord to the brain; to initiate the fine adjustments, the brain sends signals through the spinal cord and nerves to the muscle spindles to maintain muscle tone.

# Checking Out Muscle Groups

Don't *innervated, synergists,* and *antagonists* seem like great names for progressive rock bands? Maybe, but in anatomy, they're terms that apply to the muscle groups. All muscles are innervated, meaning nerves meet muscle fibers so that impulses from the nerve can cause the muscle to contract. Some innervated muscles work together — these are synergistic muscle groups. Other innervated muscles oppose each other's actions — these are antagonistic muscle groups. In this section, you find out more about the actions of muscles and why you should exercise to strengthen your muscles.

TIP

## Choose hypertrophy over atrophy

You've heard that exercise is beneficial to your health, and maybe you've even heard the saying, "Use it or lose it." When you build up muscle tissue, you not only strengthen your muscles, but you also give yourself the potential for using up the energy in your fat stores. Increased metabolism (because you're increasing your number of myofibrils) reduces the size of your fat cells. Exercising aerobically, you can use up energy in fat cells and increase the number of myofibrils in your muscles.

However, if you contract some muscles forcefully, such as when lifting weights, you increase the myofibrils in your muscles even faster and increase your metabolism. If you contract your muscles to at least 75 percent of their maximum capacity (that means lifting enough weight so that the last one or two repetitions are a bit difficult) for just a few minutes several times a week, you can experience *hypertrophy,* which is muscle tissue growth that contributes to strength and increased metabolic capability. (Don't worry ladies; women don't have the hormones to make their muscles as bulky as a man's. See Chapter 8 for more on hormones.)

If you're sedentary and your muscles only perform weak contractions, they *atrophy* — the condition in which muscle fibers shorten over time and thus weaken the muscle. When the muscle fibers decrease, the metabolic capacity of the muscle is reduced as is the strength. If you ever had an arm or leg in a cast, you probably experienced some atrophy. Fortunately, contracting your muscles through exercise can build up muscle strength and metabolism, and the benefits include

- Cancer prevention in breast, uterus, ovaries, cervix, and colon
- Decreased risk of heart attack
- Improved cholesterol levels
- Less fatigue and depression
- Less pain and swelling in arthritic joints
- Maintenance or reduction of weight, which reduces risk of type II diabetes
- Osteoporosis prevention

Exercise can be fun. If you're not one to spend an hour at the gym, you can benefit from time outdoors. Walk, ride a bike, jog, or go hiking. Play tennis or soccer, swim, ski, shoot some hoops. Just play! If you enjoy exercising with others, join a group that's into biking, running, or hiking. Join a tennis team. Go dancing. Get a dog and join a dog-walking club. Stretch your muscles and use your body. Consider it your time to play or relax and eliminate some stress. If you don't mind working out indoors, take an aerobics class, use weights and exercise tapes at home, or try yoga. Keep trying new things until you find what you like best. Choosing something that you can stick with provides the most benefit throughout your lifetime.

## *Working with you on synergists*

The word *synergy* seems overused in business these days. In my metro Atlanta phone book, almost a whole column is devoted to companies with derivatives of "synergy" — Synergy Brokerage, Synergy Films, Synergy Outdoors, Synergypaintball Enterprises (?), and Synergy Worldwide (but

I named my freelance business Synergy Publishing Services long before the word became cliché!). But synergy really is a medical word that means working together. And — if you don't know it yet, you will soon — many parts of the body work together to achieve a common goal.

In the muscle system, groups of muscles often work together to move a body part. Muscles that work this way are said to be synergistic. The muscle that does most of the moving is the *prime mover;* the muscles that help the prime mover achieve a certain body movement are *synergists*.

## Opposing each other: Antagonists

Antagonistic muscles oppose each other, but in doing so, they achieve an effect together. One example is flexing your arm. When you bend your forearm up toward your shoulder, your biceps muscle contracts, but the triceps muscle in the back of your arm relaxes. The actions of the biceps and triceps muscles is opposite, but both actions allow the arm to flex. Then when you want to lower your arm, antagonistic actions allow you to do so: The biceps muscle relaxes, and the triceps muscle contracts.

# Locating Skeletal Muscles of the Body

Okay. The muscles that your body uses to hold itself together and move around form half of the musculoskeletal system, which is the combination of bones and muscles that compose your body. The musculoskeletal system forms your body's "shell," the "container" that your organs, nerves, and blood vessels fill. This section focuses on the skeletal muscles and starts with how these muscles are named.

## Making muscle sense

To name muscles, anatomists had to come up with a set of rules to follow so that the names would make sense. They chose to focus on certain characteristics from which to derive the Latin name of a muscle. If necessary, refer back to Chapter 1 for information on anatomic position. Examples of characteristics in muscle names are given in Table 5-1.

| Table 5-1 | Characteristics in Muscle Names |
|---|---|
| *Characteristics* | *Examples* |
| Muscle size | The largest muscle in the buttocks is the *gluteus maximus* (*maximus* means large in Latin); a smaller muscle in the buttocks is the *gluteus minimus* (*minimus* means small in Latin). |
| Muscle location | The *frontalis muscle* lies on top of the skull's frontal bone. |
| Muscle shape | The *deltoid muscle,* shaped like a triangle, comes from *delta* — the Greek alphabet's 4th letter, which is also shaped like a triangle. |
| Muscle action | The *extensor digitorum* is a muscle that extends the fingers or *digits.* |
| Number of muscle attachments | The *biceps brachii* attaches to bone in two locations, whereas the *triceps brachii* attaches to bone in three locations. |
| Muscle fiber direction | The *rectus abdominis muscle* runs vertically along your abdomen (*rectus* means straight in Latin). |

Many of the terms used in anatomy and physiology (as well as medicine and other sciences) are derived from Latin and Greek words. So if you can pick apart a term into its roots and stems, your knowledge blossoms. Like all the floral analogies? It's true. Remembering that *bi* means "two," *tri* means "three," *max* means "large," and *brach* means "arm" can help you remember other details about anatomic parts. The best way to get a firm grasp on these roots and stems is to keep reading the subject matter. The more of it that you read, the more of it sinks in and becomes easier to recall. Trust me. I've been in the shoes that cover your *flexor longus digitorum* (a long muscle in the leg and sole of the foot that allows your toes to flex).

## Naming muscles from top to bottom

Get ready because I'm telling you the muscle names from head to toe, literally. Yes, even your toes have muscles.

### *Starting at the top*

Your head contains muscles that perform three basic functions: chewing, making facial expressions, and moving your neck. I guess ear wiggling falls into this category, too.

To chew, you use the muscles of *mastication* (a big, fancy word meaning "chewing"). The *masseter,* a muscle that runs from the zygomatic arch (your cheekbone) to the mandible (your lower jaw), is the prime mover for mastication, so its name is based on its action (*mass*eter, *mas*tication) The fan-shaped *temporalis muscle* is a synergist to the masseter. The temporalis muscle works with the masseter to allow you to open and close your jaw and lies on top of the skull's temporal bone, so its name is based on its location. Figure 5-2 shows the muscles of the head and neck.

To smile, frown, or make a funny face, you use several muscles. The *frontalis muscle* (see Table 5-1) along with the a tiny muscle called the *corrugator supercilii* raises your eyebrows and gives you a worried or angry look when wrinkling your brow. (Think of the appearance of corrugated cardboard, and then feel the skin between your eyebrows when you "wrinkle your brow.") The *orbicularis oculi muscle* surrounds the eye (the word *orbit,* as in *orbicularis,* means to encircle; *oculi* refers to the eye). This muscle allows you to blink your eyes and close your eyelids, but it also gives you those little crow's feet at the corners of your eyes. The *orbicularis oris* surrounds the mouth. (*Or* refers to mouth as in "oral.") You use this muscle to pucker up for a kiss.

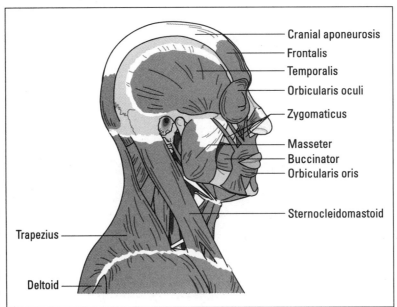

**Figure 5-2:**
The muscles of the neck and head.

From LifeART®, Super Anatomy 1, © 2002, Lippincott Williams & Wilkins

If you play the trumpet or another instrument that requires you to blow out, you're well aware of what your *buccinator muscle* does. This muscle is in your cheek. (*Bucc-* means cheek, as in the word buccal, which refers to the cheek area.) It allows you to whistle and also helps keep food in contact with your teeth as you chew. Remember that your zygomatic arch is your cheekbone? Well, the *zygomaticus muscle* is a branched muscle that runs from the cheekbone to the corners of your mouth. This muscle pulls your mouth up into a smile when the mood strikes you. Figure 5-3 shows the facial muscles. Don't forget to look for the branched zygomaticus muscle.

When you want to nod *yes, no,* or tilt your head into a *maybe so,* your neck muscles come into play. You have two *sternocleidomastoid muscles,* one on each side of your neck. I know this is a long name, but the name reflects the locations of its attachments: the sternum, collar bone, and mastoid process of the skull's temporal bone. When both sternocleidomastoid muscles contract, you can bring your head down toward your chest and flex your neck. When you turn your head to the side, one sternocleidomastoid muscle contracts (the one on the opposite side of the direction your head is turned). So if you turn your head to the left, your right sternocleidomastoid muscle contracts and vice versa. If you want to lean your head back, like to look up at the sky, or shrug your shoulder, your *trapezius muscle* allows you to do so.

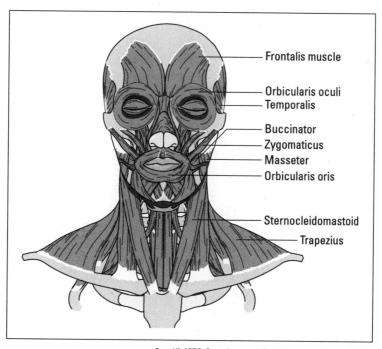

**Figure 5-3:**
The
muscles of
the face.

Frontalis muscle

Orbicularis oculi
Temporalis

Buccinator
Zygomaticus
Masseter
Orbicularis oris

Sternocleidomastoid
Trapezius

*From LifeART®, Super Anatomy 1, © 2002, Lippincott Williams & Wilkins*

The trapezius is an antagonist to the sternocleidomastoid muscle. If you remember basic geometry, a trapezoid is shaped like a diamond, and that's exactly what the trapezius looks like. It runs from the base of your skull to your thoracic vertebrae and connects to your shoulder blades. Therefore, the trapezius and sternocleidomastoid muscles connect your head to your torso and provide me with a nice segue to the next section. Take a look at Figure 5-4 to see a rear view of the neck and torso muscles.

### Twisting the torso

The torso muscles have important functions. They not only give support to your body, but they also connect to your limbs to allow movement, allow you to inhale and exhale, and protect your internal organs. In this section, I'll cover the muscles that run along the front of you (called your anterior or ventral side) and then cover the muscles of your back (your posterior or dorsal side).

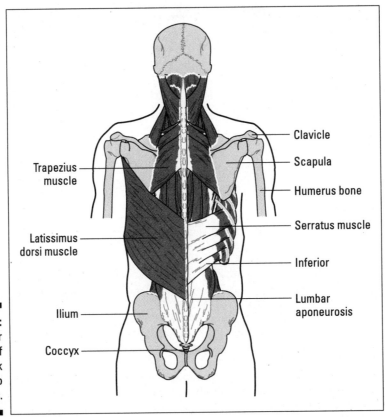

**Figure 5-4:**
Posterior
view of
the neck
and torso
muscles.

Trapezius
muscle

Latissimus
dorsi muscle

Ilium

Coccyx

Clavicle

Scapula

Humerus bone

Serratus muscle

Inferior

Lumbar
aponeurosis

*From LifeART®, Super Anatomy 1, © 2002, Lippincott Williams & Wilkins*

In your chest (see Figure 5-5), your *pectoralis major muscles* connect your torso at the sternum and collar bones to your upper limbs at the humerus bone in the upper arm. Your "pecs" also help to protect your ribs, heart, and lungs. You can feel your pectoralis major muscle working when you move your arm across your chest. Also in your chest are the muscles between and around the ribs. The *internal intercostal muscles* help to raise and lower your rib cage as you breathe. However, the torso's largest muscles are the abdominal muscles.

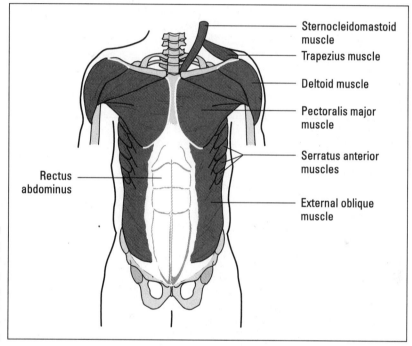

Sternocleidomastoid muscle

Trapezius muscle

Deltoid muscle

Pectoralis major muscle

Serratus anterior muscles

Rectus abdominus

External oblique muscle

**Figure 5-5:** Anterior muscles of the chest and abdomen.

From LifeART®, Super Anatomy 1, © 2002, Lippincott Williams & Wilkins

The abdominal muscles really form the center of your body. If the abdominal muscles are weak, the back is weak because the abdominal muscles help to flex the vertebral column. So if the vertebral column doesn't flex easily, the muscles attached to it can become strained and weak. And the muscles of the abdomen and back join to the upper and lower limbs. Therefore, if the abdomen and back are weak, the limbs can have problems.

The muscles of the abdomen are thin, but the fact that the muscle fibers of the abdominal muscles run in different directions increases their strength. This woven effect makes the tissues much stronger than they would be if they all went in the same direction. As I write this, my daughter asks for help connecting some building blocks. While showing her that laying a top layer of blocks perpendicular to the blocks underneath helps the structure stay together, it occurs to me that this is similar to how the abdominal muscle tissues provide strength and stability. Hey, blocks inspire my kids to be creative, why not me, too?

The "washboard" muscle of the abdomen, the *rectus abdominis,* forms the front layer of the abdominal muscles, and it runs from the pubic bones up to the ribs and sternum. The "job" of the rectus abdominis muscle is to hold in the organs of the abdomino-pelvic cavity and allow the vertebral column to flex.

Other layers of abdominal muscles also help to hold in your organs on the side of your abdomen and provide strength to the core of your body. The *external oblique* muscle attaches to the eight lower ribs and runs downward toward the middle of your body (slanting toward the pelvis). The *internal oblique muscles* lie underneath the external oblique (makes sense, eh?) at right angles to the external oblique muscles. The internal oblique muscles extend from the top of the hip at the iliac crest to the lower ribs.

Together, the external and internal oblique muscles form an X, essentially strapping together the abdomen. The abdomen's deepest muscle, the *transversus abdominis,* runs horizontally across the abdomen; its function is to tighten the abdominal wall, push the diaphragm upward to help with breathing, and help the body bend forward. The transversus abdominis is connected to the lower ribs and lumbar vertebrae and wraps around to the pubic crest and *linea alba.* The linea alba ("white line") is a band of connective tissue that runs vertically down the front of the abdomen from the xiphoid process at the bottom of the sternum to the pubic symphysis (the strip of connective tissue that joins the hip bones).

The muscles in your back (refer to Figure 5-4) serve to provide strength, join your torso to your upper and lower limbs, and protect organs that lie toward the back of your trunk (such as your kidneys). The *deltoid muscle* joins the shoulder to the collar bone, scapula, and humerus. This muscle is shaped like a triangle — think of the Greek letter delta, which looks like this: Δ The deltoid muscle helps you to raise your arm up to the side (that is, laterally). The *latissimus dorsi muscle* is a wide muscle that's also shaped like a triangle. It originates at the lower part of the spine (thoracic and lumbar vertebrae) and runs upward on a slant to the humerus. Your "lats" allow you to move your arm down if you have it raised and to reach, such as when you are climbing or swimming.

### Spreading your wings

Your upper limbs — namely, your arms — have a wide range of motions. Obviously, your arms are connected to your torso. One of the muscles that provides that connection, the *serratus anterior,* is below your armpit (the anatomic term for armpit is *axilla*) and on the side of your chest. The serratus anterior muscle connects to the scapula and the upper ribs. You use this muscle when you push something or raise the arm higher than horizontal. Its action pulls the scapula downward and forward.

Although the *biceps brachii* and *triceps brachii* are muscles located in the top (anterior) part of your upper arm, their actions allow the forearm (lower arm) to move. Figure 5-6a shows an anterior view of the upper limb. You can feel your biceps muscle move when you pretend to turn a doorknob and rotate your forearm. The name *bi*ceps refers to this muscle's two origins (points of attachment); it attaches to the scapula in two places. From there, it runs to the radius of the forearm (its point of insertion). The triceps brachii is the only muscle that runs along the back (posterior) side of the upper arm. Figure 5-6b shows a posterior view of the upper limb. The name *tri*ceps refers to the fact that it has three attachments: one on the scapula and two on the humerus. It runs to the ulna of the forearm. You can feel this muscle in motion when you push or punch. Other muscles of the arm include the *brachioradialis,* which helps to flex the arm at the elbow, and the *supinator,* which rotates your arm from a palm-up position to a palm-down position.

The hand contains muscles that move the whole wrist and hand, as well as muscles that perform the fine movements of the fingers. When you type or play the piano, you're using your *extensor digitorum* and *flexor digitorum muscles* to raise and lower your fingers onto the keyboard and move them to the different rows of keys. As you lift your hands off the keyboard, the muscles of your wrist kick into gear. The *flexor carpi radialis* (attached to the radius bone) and *flexor carpi ulnaris* (attached to the ulna bone) allow your wrist to flex upward. The *extensor carpi radialis longus* (which passes by the carpal bones), the *extensor carpi radialis brevis,* and the *extensor carpi ulnaris* allow the wrist to extend; that is, bend forward/downward.

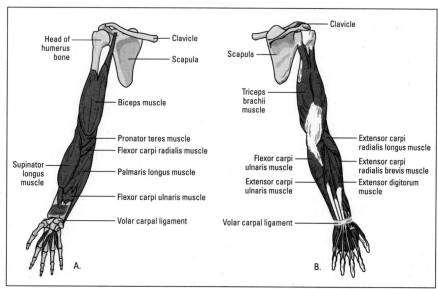

**Figure 5-6:**
The muscles of the upper limb: Anterior (A) and posterior (B).

*From LifeART®, Super Anatomy 1, © 2002, Lippincott Williams & Wilkins*

# Thumbs up!

A key feature of all primates is a *prehensile thumb;* that is, a thumb that's adapted for grasping objects. Many animals have digitlike structures, but only primates can grasp things with their hands. And the only way to grasp things is to have a thumb. Imagine having webbing between your four fingers so that you couldn't spread them apart; you wouldn't be able to pick things up. That's why animals, such as dogs, cats, and birds, hold things in their mouths (or beaks). But, primates — apes, monkeys, and humans — can easily grasp things between

their thumb and fingers. However, of those primates, only humans have an opposable thumb (one that can touch each of the other fingers; the thumb can be "opposite" from each finger). Because of the ability to oppose the thumb to each finger, the muscles in our digits are capable of performing minute movements. As you touch your thumb to your pinky finger, your palm becomes arched, which only happens in humans because of short bones in the pinky and an opposable thumb.

### Getting a leg up

Your lower limbs consist of your legs, which are connected to your buttocks. Your buttocks are connected to your hips. So consider your lower limb to be what is shown in Figure 5-7.

The *iliopsoas muscle* connects your lower limb to your torso and consists of two smaller muscles: the *psoas major,* which joins the thigh to the vertebral column, and the *iliacus,* which joins the hipbone's ilius to the thigh's femur bone. Originating on the iliac spine of the hip and joining to the inside surface of the tibia (a bone in your shin), the *sartorius muscle* is a long, thin muscle that runs from the hip to the inside of the knee. These muscles stabilize the lower limbs and provide strength for the legs to support the body's weight and balance the body against the pressure of gravity.

Some muscles in the leg allow the thigh to move in a variety of positions. The buttocks muscles allow you to straighten your leg at the hip and extend your thigh when you walk, climb, or jump. The *gluteus maximus* — the largest muscle in the buttocks — is the largest muscle in the body (see Figure 5-8). The gluteus maximus is antagonistic to the stabilizing iliopsoas muscle, which flexes your thigh. The *gluteus medius muscle,* which lies behind the gluteus maximus, allows you to raise your leg to the side so that you can form a 90-degree angle with your two legs (this action is *abduction* of the thigh). Several muscles serve as adductors; that is, they move an abducted thigh back downward. These muscles include the *pectineus* and *adductor longus,* which become injured when you "pull a groin muscle," as well as the *adductor magnus* and *gracilis,* which run along the inside of your thigh.

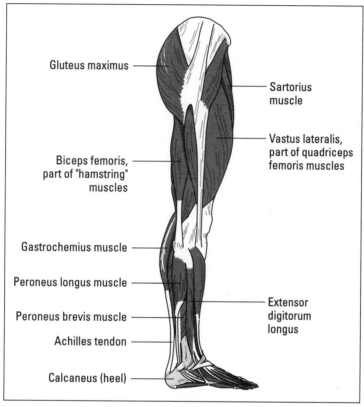

- Gluteus maximus
- Sartorius muscle
- Vastus lateralis, part of quadriceps femoris muscles
- Biceps femoris, part of "hamstring" muscles
- Gastrochemius muscle
- Peroneus longus muscle
- Peroneus brevis muscle
- Extensor digitorum longus
- Achilles tendon
- Calcaneus (heel)

**Figure 5-7:**
The muscles of the lower limb.

*From LifeART®, Super Anatomy 1, © 2002, Lippincott Williams & Wilkins*

# Where did these names come from?

Some muscle names have a pretty interesting history. Take the hamstring muscles and the sartorius muscle. First the hamstrings — ham may make you think of pigs, and, yes, pigs have hamstrings in their legs. And the biceps femoris, semimembranosus, and semitendinosus muscles have the same strong tendons in a pig as they do in you. When butchers smoked hams (thigh meat from a pig), they hung the hams on hooks in the smokehouse by these ropelike tendons, which generated the name "hamstrings."

Nobody said butchers were creative. (Don't be offended, butchers. I come from a long "string" of butchers.)

The sartorius muscle goes into action when you sit cross-legged, like tailors used to do when they pinned hems or cuffs (and maybe still do). So the sartorius muscle is sometimes referred to as the tailor's muscle. And guess what means "tailor" in Latin? Yep, *sartor*.

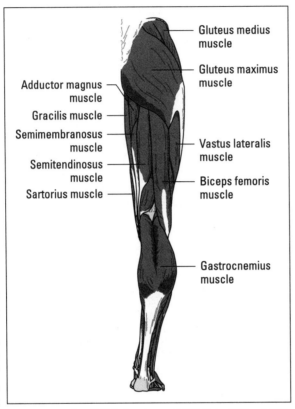

Gluteus medius muscle

Gluteus maximus muscle

Adductor magnus muscle

Gracilis muscle

Semimembranosus muscle

Semitendinosus muscle

Sartorius muscle

Vastus lateralis muscle

Biceps femoris muscle

Gastrocnemius muscle

**Figure 5-8:** The muscles of the posterior lower limb.

From LifeART®, Super Anatomy 1, © 2002, Lippincott Williams & Wilkins

Other muscles in the thigh serve to move the lower leg. Along your thigh's front and lateral side, four muscles work together to allow you to kick. These four muscles — the *rectus femoris, vastus lateralis, vastus medialis,* and *vastus intermedius* — are better known as the *quadriceps (quadriceps femoris). Quad* means four as in quadrilateral or quadrant. Refer to Figure 5-7 for a look at the lateral side of the lower limb; Figure 5-9 shows the muscles of the anterior lower limb.

The hamstrings are a group of muscles that are antagonistic to the quadriceps. The hamstrings — the *biceps femoris, semimembranosus,* and *semitendinosus* — run down the back of the thigh (refer to Figure 5-8) and allow you to flex your lower leg and extend your hip. They originate on the ischium of the hipbone and join (insert) to the tibia of the lower leg. You can feel the tendons of the hamstring muscles behind your knee.

Your lower leg's shin and calf muscles move the ankle and foot. The *gastroc-nemius,* better known as the "calf muscle," begins (originates) at the femur (thigh bone) and joins (inserts at) the Achilles tendon that runs behind your heel. You can feel your gastrocnemius muscle contracting when you stand on your toes. The antagonist of the gastrocnemius, the *tibialis anterior,* starts on the surface of the tibia (shin bone), runs along the shin, and connects to the ankle's metatarsal bones. You can feel this muscle contract when you raise your toes and keep your heel on the floor. The *peroneus longus* and *peroneus brevis* (*brevis* meaning short as in "brevity") run along the outside of the lower leg and join the fibula to the ankle bones. In doing so, the peroneus muscles help to move the foot. The *extensor digitorum longus* and the *flexor digitorum longus muscles* join the tibia to the feet and allow you to extend and flex your toes, respectively, like the fingers.

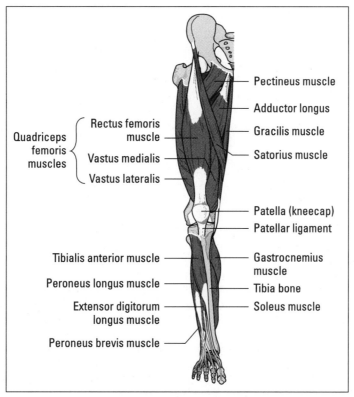

**Figure 5-9:**
The
muscles of
the anterior
lower limb.

Pectineus muscle

Adductor longus

Gracilis muscle

Satorius muscle

Quadriceps femoris muscles
Rectus femoris muscle
Vastus medialis
Vastus lateralis

Patella (kneecap)

Patellar ligament

Tibialis anterior muscle

Peroneus longus muscle

Extensor digitorum longus muscle

Peroneus brevis muscle

Gastrocnemius muscle

Tibia bone

Soleus muscle

*From LifeART®, Super Anatomy 1, © 2002, Lippincott Williams & Wilkins*

# *Pathophysiology of the Muscular System*

Every system of the body has its share of things that can go wrong. The muscular system is no exception. Muscles can spasm, causing pain and limiting movement. Requiring much rehabilitation to become fully repaired, tendons and ligaments can tear. Disease can affect the muscles, too.

## *Muscle spasms*

Ouch. I have experienced many of these, and they do have a knack for generating pain and difficulty in moving certain body parts. I'm not a hyper person, but I rarely relax. Unfortunately, my muscles often don't relax either. Occasionally, I pay the price.

A few sections ago, I told you about the gluteus maximus and gluteus medius in the buttocks. Well, another little muscle in the buttocks, the *piriformis,* can cause a "pain in the butt." The piriformis muscle, a tiny, stringlike muscle that originates from the ilium of the hip and the sacrum, runs to the upper border of the greater trochanter, a point at the top of the femur. When the piriformis muscle works properly, it rotates the thigh out to the side (laterally). However, when it's bad, it's very, very bad.

A *muscle spasm* is a sudden, involuntary contraction. Causing sudden pain, the muscle tenses up violently and without warning. When my little piriformis muscle undergoes a spasm, the pain is unbelievable. Being attached to the sacrum at the base of the spine, a piriformis muscle spasm causes pain in the lower back; afterward, the buttock and sometimes the hip (the piriformis muscle is also attached to the hip) feel bruised. The pain is noticeable and the spasm is limiting because when it contracts so forcefully, it can irritate the sacral nerves.

The piriformis muscle spasm is my personal experience. But, of course, a spasm can occur in any muscle, and the effects vary according to the location and nerves that are nearby. Not all spasms are painful. Hiccups, which are the result of spasm in the diaphragm, usually are not painful — annoying but not painful. Muscle cramps are spasms, too. The calf muscle (gastrocnemius) is a common place for sudden cramps to occur. Facial tics, such as when your eyelid twitches repeatedly, also are muscle spasms, and they can be just as annoying as hiccups, right?

# Muscular dystrophy

*Muscular dystrophy* is an inherited, chromosomal disorder, but it affects the muscles.

*Duchenne muscular dystrophy* (DMD) is the most common type, but others exist. Like I said, muscular dystrophy is an inherited disorder. Most often, DMD is passed from mother to son. Boys are most commonly affected with DMD, and the symptoms become evident usually before the boy is 3 years old. What happens to these boys is that their muscles slowly weaken, shorten, and degenerate. As the disease progresses, their muscles eventually waste away. The boys end up disabled and confined to a wheelchair at about age 12. Fat and connective tissue replace normal muscle tissue, thus causing problems in the heart and lungs. DMD patients usually die when they're teenagers.

*Myotonic muscular dystrophy* can affect males or females and can start at different ages (variable age of onset). These patients experience increasing muscle weakness and stiffness. Eventually, problems during actions, such as swallowing, occur because muscles don't relax after contractions. The muscles in the face and neck are usually the first to show signs of being affected; then, the arms and legs become affected. Turning the head or lifting objects becomes difficult; a person affected with myotonic muscular dystrophy progressively worsens and often becomes confined to a wheelchair or bed.

# Chapter 6

# The Great Coverup: Skin

*H*ave you ever thought of your skin as an organ? Well, it is. As you read this chapter, you'll understand why the skin is classified as an organ, and you'll take a look at the physiological processes that occur in this large, thin organ. This chapter shows you what structures lie beneath your skin and what diseases affect the skin. The pathophysiology of some skin disorders is explained, including burns. Without your skin, you would dehydrate, and bacteria and viruses — commonly called germs (see Chapter 13) — would inundate your body. In short, you couldn't survive. So show your skin some respect. Find out what it's made of and what it does for your body on a daily basis. And slather on some sunscreen. Okay?

## Getting Triple Protection

You're wrapped in roughly 21 square feet of skin, also known as *integument* or *cutaneous membrane*. Figure 6-1 shows the anatomy of the skin. And although your skin feels tight, it really is loosely attached to the layer of muscles below. In spots where muscles don't exist, such as on your knuckles, the skin is attached right to the bone.

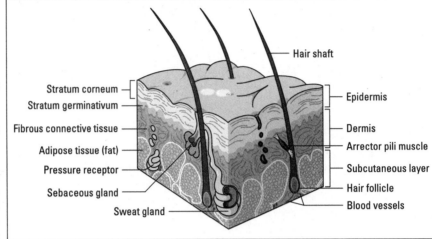

**Figure 6-1:**
Cross section of the skin showing the layers and some specialized structures.

Hair shaft

Stratum corneum

Stratum germinativum

Fibrous connective tissue

Adipose tissue (fat)

Pressure receptor

Sebaceous gland

Sweat gland

Epidermis

Dermis

Arrector pili muscle

Subcutaneous layer

Hair follicle

Blood vessels

*From LifeART®, Super Anatomy 1, © 2002, Lippincott Williams & Wilkins*

The skin has two main layers: *epidermis* and *dermis,* plus the *subcutaneous layer* that lies underneath the dermis. The subcutaneous layer serves as a go-between for the outer layers of skin and the muscles and organs that are underneath the skin. Because the skin contains some other structures (such as blood vessels), the skin can be classified as an organ system — the *integumentary system.* The skin, your body's largest organ, accounts for approximately 15 percent of your body weight.

## Touching the epidermis

The epidermis is the part of your skin that you can touch. This thin, outer layer is made up of two layers: the *stratum corneum* and the *stratum germinativum.* All of the epidermis is composed of stratified squamous epithelium tissue (see Chapter 3), but the two regions of the epidermis perform different functions.

Stratum, by the way, is a Latin-derived term for layer meaning "bed" or "spread out" (think stratosphere).

### Keratinizing the stratum corneum

The stratum corneum is the tough, outer layer that's 25 to 30 cells thick. What makes the layer tough is that the cells in this layer become hardened through *keratinization* — a process in which the cells produce the protein keratin. Many keratin fibers harden during this process and provide strength.

Forming a barrier between the external environment and your inner cells, keratin is waterproof. Keratin protects you from losing water to the outside or gaining water to the inside, so you don't dehydrate or swell.

The hardened keratin cells protect your inner cells and tissues from bacteria and viruses. Thus any openings on the skin (such as cuts or scrapes) must be cleaned out well. Otherwise, the microbes have access to the inside of your body where they can cause infection.

### Farming the stratum germinativum

The stratum germinativum (also called the *stratum basale* or *basal layer*) is like a cell farm. This lower region of the epidermis (refer to Figure 6-1) contains cells that are constantly dividing to produce new cells. The new cells rise up to the surface (the stratum corneum) in 14 to 30 days, but as they rise, they move farther and farther away from the blood supply and die within a short time. So the skin cells that you show the world are actually dead cells. And every day, some of the dead cells slough off.

Don't let the bedbugs bite! Your mattress is full of your dead skin cells, and those dead skin cells are extremely appetizing to bedbugs, which are tiny mites that decompose the cells but leave their excrement behind. Some people are highly allergic to these little buggers.

The stratum germinativum also contains *Langerhans' cells* — special cells that connect the integumentary system and the immune system. *Langerhans' cells* are made in the red bone marrow but settle in the stratum germinativum. The Langerhans' cells eat (phagocytize) bacteria and viruses and then carry them through the bloodstream to organs of the lymphatic system (such as the spleen). When phagocytized microbes show up in a lymphoid organ, the immune system is stimulated to actively pursue similar microbes within your body to prevent a major infection from occurring. (See Chapter 13 for more information on microbes, immunity, and the lymphatic system and its organs.)

Another important function of the stratum germinativum is to provide your skin with enough color to help protect you from the sun. *Melanin* — the skin pigment compound — is produced in *melanocytes* that reside in the stratum germinativum. The purpose of melanin is to protect the skin from the sun's ultraviolet rays. The more melanin that a person produces, the darker their skin looks. So races that dominated geographic areas close to the equator — where the sun's rays are most intense — produce more melanin because these people had more of a need for protection from the sun's ultraviolet rays. Racial groups that dominated geographic areas far from the equator produce little melanin because they had less of a need for that protection.

Your skin pigmentation helps to protect you from the sun's damaging ultraviolet rays.

Although everybody has about the same number of melanocytes, an individual's genes dictate the level of melanin production that results in skin color differences. In addition, people of Mediterranean descent usually have more of a greenish pigment mixed with their melanin, and people of Asian descent have a yellowish pigment mixed with their melanin. But everybody has about the same number of melanocytes, and underneath the stratum germinativum, everybody looks the same.

## Meddling in the middle: The dermis

Like a middle child, the dermis is sandwiched between the epidermis, which gets all the attention (lotion and such), and the subcutaneous layer, which people try to keep thin. (See more on the subcutaneous layer in the next section.) But like middle children, the thick dermis has some important responsibilities of its own.

The dermis contains collagen and elastin fibers. Collagen is a huge protein, and fibers that are made of it are extremely stretchy and flexible as are the elastin fibers that provide the elasticity for your skin to stretch during movements. Collagen prevents the skin from overstretching and tearing during movements. Can you imagine ripping your skin like a pair of pants when you bend over? What if your skin tore like the armpit of an old shirt when you reached for something high on a shelf?

Blood vessels that provide the skin cells with oxygen and nutrients are contained in the dermis. When you exercise or are embarrassed to the point of blushing, the blood vessels dilate (open) and allow more blood to flow into the skin. The increased number of red blood cells near the body's surface creates the redness on your face when you blush. During exercise, the dilation of the blood vessels allows heat generated to escape from the body in an attempt to keep body temperature normal. When you are cold, the blood vessels constrict in an attempt to retain body heat.

By having fewer red blood cells near the skin surface, body parts, such as your lips, turn bluish. When the bluish color results from a physiological problem, such as lack of oxygen in the blood or interruption of blood flow, it's referred to as *cyanosis*. If you ever looked at a color cartridge for a printer, you may have noticed that the blue is called *cyan* — the root in several words: *cyanobacterium* are blue-green algae; and, *cyanide* is colorless, but a person who consumes it develops cyanosis just before dying.

Your dermis also provides each person with fingerprints and footprints. Yes, the fingerprints and footprints are visible on the outer layer — the epidermis — but the way that the epidermis connects to the dermis creates the ridges of a fingerprint. High in the dermal layer are projections called *dermal papillae* that in essence pull the epidermis down toward the dermis. So ridges result in the anchored epidermis. Even the ridges have a function: They allow you to grip really tiny objects better because friction is increased between the surface of an object and the *epidermal ridges*. Everyone has these ridges, but the location of the dermal papillae underneath the epidermis varies from person to person. Everybody's fingertips and toes have unique print patterns.

The dermis serves as the location for special receptors in the skin. These receptors help you feel sensations, such as pressure, vibration, and light touch. The receptors with names, such as *Pacinian corpuscles, Meissner's corpuscles,* and *Ruffini's corpuscles,* are sprinkled throughout the dermis and are connected to the nerves that run through the dermis and subcutaneous layer. At one spot on your skin, you may sense light touch, while a few centimeters away, you may sense pressure. Not every inch of skin is covered with receptors for every sensation. To find out more about how the nerves transmit sensations received in the skin's receptors, see Chapter 7.

# Getting under your skin: The subcutaneous layer

The subcutaneous layer is also called the *hypodermis* because it lies directly beneath the dermis (*hypo-* means below, lower than). This term may sound familiar if you've ever heard of hypodermic needles. Hypodermic needles are used to reach your subcutaneous layer.

The size of the subcutaneous layer can vary greatly for the simple reason that this is where "fat cells" exist. Fat is deposited into the cells of the *adipose tissue,* which makes up the majority of the subcutaneous layer. So really skinny people have a thin subcutaneous layer, and obese people have a thick subcutaneous layer.

Not all fat is bad, though. You need some fat to be healthy. Adipose tissue insulates the body and protects the organs lying underneath the subcutaneous layer when you fall. Fat also provides a storage place for energy, and it literally hangs around until you need it.

# Accessorizing Your Skin

Sorry. This section has nothing to do with tattoos or body piercing. It's all about your skin's *accessory structures:* hair, nails, and glands — structures that work with the skin.

## Now hair this

Your body has millions of hair follicles — about the same number as on the closest relative to humans, the chimpanzee. (Although I don't know who counted all of them!) And like a chimpanzee, humans have hairless palms, soles, lips, and nipples. Most of your hair is lightweight, fine, and downy. But the hair on your head is heavier to help hold in body heat. Puberty brings about a surge of sex hormones that stimulate hair growth in the axillary (armpit) and pelvic regions. Men also grow facial hair, and women with hormonal imbalances can develop facial hair, too (see Chapter 8).

All hairs stick out of hair follicles — small tubes made up of epidermal cells that extend down into the dermis. They extend into the dermis so that the hair root is supplied with blood. Although the hair outside the skin is dead, the hair follicle produces new cells that push the dead hair out further (producing "growth"). So, just like the root of a plant needs to extend down into the soil to get water and minerals, your hairs need to extend down into your scalp, face, arm, leg, chest, or groin. Inside the dermis of hair-covered skin, the hair root adjoins cells in the hair follicle that supply needed oxygen and nutrients, which are delivered through the bloodstream. Think of the hair on your head as a nice patch of lawn. (If you're balding, you just have a little more landscaping and less lawn to mow than others!)

Cells at the bottom of the hair follicle continually divide to produce new cells. Those new cells form a hair. When the hair is new and short enough to be down near the blood vessels in the dermis, it's well nourished and alive. But, like other skin cells, the hair rises up through the layers as it grows. On its way up and out, the hair cells get farther away from the blood supply, then they become keratinized, and eventually, they die. Like the skin on your body, the hair you see is made of dead cells. And, like skin, they shed.

The hairs on your head live about three to four years before being shed; eyelashes live about three to four months before falling out and being replaced. People don't go bald overnight. Baldness (called *alopecia*) occurs when hairs are no longer replaced.

# Enough to make your hair stand on end

Each hair follicle has a tiny *arrector pili* muscle. When these smooth muscles contract, a goose bump appears. When you're cold or frightened, the sudden tightening of the hair follicle traps air between the hair and the skin and causes the hair to stand on end. When you're cold, the trapped air acts as insulation. When you're frightened, the hairs standing on end serve to make you look scary to whatever is scaring you. Remember, this physiological response began millions of years ago when humans were both predator and prey. These days, you may get goose bumps while watching a scary movie; the fear within your brain may be the same, but your body doesn't know that the movie screen isn't a predator.

## Growing nails

Your fingernails and toenails lie on a nail bed (not to be confused with a bed of nails). At the back of the nail bed is the nail root. Just like skin and hair, nails start growing near the blood supply that lies under the nail bed, and the cells move outward. As they move out over the nail bed, they become keratinized. Keratinization can make a tissue harder as in the case of nails, or waterproof, as in the case of skin and hair. And, yes, skin and hair get wet, but they don't absorb water; they shed it because of keratin. The nails that you see on your hands and feet are the old, keratinized cells that form the nail body. The edge of your nails that appears to grow doesn't really grow; it just gets pushed farther out as new cells are created. Your nail "grows" about one millimeter in one week.

At the bottom of your nails is a white, half-moon shaped area called the *lunula*. (*Lun*- is the Latin root for moon as in *lunar*.) The lunula is white because this is the area of cell growth. In the nail body, the nail appears pink because the blood vessels lie underneath the nail bed. But many more cells fill in the area of growth. This layer is thicker, and you see white instead of pink.

If the *cuticle* — the skin that covers the nail at the bottom of the lunula — becomes dry and cracked, germs can get into the nail bed and thus cause infection. Sometimes, fungi get into the nail bed and then cause a fungal infection in the nails. Biting your cuticles or nails can also lead to infection because bacteria from your mouth are able to get to the vascularized (filled with blood vessels) tissue alongside or underneath the nail.

Nails can give doctors clues about problems in the body. Table 6-1 shows you some nail problems and the disease or disorder that may be affecting the person.

| Table 6-1 | Nail Problems as Signs of Possible Medical Conditions. |
|---|---|
| *Nail problem* | *Indicated Disease or Disorder* |
| Brittle, concave (spoon-shaped) nails with ridges | Iron-deficiency anemia |
| Nails separating from the nail bed | Thyroid disorder in which too much thyroid hormone is produced (such as Graves' disease) |
| Black marks (look like splinters) under the nails | Respiratory disease or heart disease |
| Hard, curved, yellow nails | Bronchiectasis (chronic dilation of bronchial tubes, bad breath, spasmodic coughing) and lymphedema (fluid backing up in lymph glands that causes swelling of hands and feet). Smoking cigarettes and wearing nail polish can also stain the nails yellow, too. |

# Nothing's bland about glands

The body's glands produce substances, and then the glands export them for delivery. Glands near the internal organs produce substances that have an effect somewhere else within the body. For instance, the pancreas, a large gland associated with the stomach, produces insulin in response to how much glucose you take in. The bloodstream picks up the insulin and keeps your glucose level within the normal range. However, in the skin, glands secrete substances that go outside of the body.

The two main types of skin glands are *sweat glands* (called *sudoriferous glands*) and *sebaceous glands*.

### Sweating it out

Yes, sweat glands make *sweat* — a watery substance that serves as a way for the body to remove excess salt (sodium chloride) and urea (the waste product excreted by the kidney), as well as being extremely important in regulating body temperature. (For more on the kidneys, see Chapter 12.)

✔ **Eccrine sweat glands:** Distributed all over the body, these sweat glands open to the skin's surface when you're hot and let heat escape to reduce body temperature. The way that heat escapes the body is described by the process of *evaporative cooling*. Sweat is a watery substance. Excess body heat is absorbed by the water in sweat. As the watery sweat evaporates, the heat is moved away from the body, thus cooling it.

✔ **Apocrine sweat glands:** Responsible for body odor, these glands are found deep in the hair follicles of the armpits and groin, and they start to develop during puberty. Whereas sweat from the eccrine glands is clear, sweat from the apocrine glands contains a milky white substance. Bacteria on the skin that digest the milky white substance produce byproducts that, well, smell. Apocrine glands become active when you're anxious and stressed as well as when you're sexually stimulated.

### Secreting healthy oil

Sebaceous glands secrete an oily substance called *sebum* into hair follicles. Besides wreaking havoc with teenage facial pores, sebum has physiological functions. The less hair you have, the faster you lose body heat, so sebum helps take care of your hair. The oily nature of sebum keeps hair soft so that it doesn't break easily, and like keratin, it's another substance that keeps epidermal cells "waterproof." Every year my husband waterproofs his boots with mink oil; the oil keeps water from getting into the leather of the boots. Sebum is your body's waterproofing agent. When the sebaceous glands attached to hair follicles secrete sebum, the sebum coats the skin cells and prevents water loss to the outside as well as keeping water from getting inside your body. Making the skin surface an inhospitable place for some bacteria, sebum also helps to protect you from infection.

# Saving Your Skin

As an organ, your skin performs specific functions. Just like your heart, lungs, and brain, it has things to do, and your skin must do them well enough to keep you healthy. The skin acts like a brave soldier. It puts itself out and gets hit with ultraviolet rays, pain, pressure, heat, and cold. The body's frontline defense, your skin must heal itself after being wounded. Okay, men and women, read on to find out what your skin does. And be careful out there!

## Making vitamin D

Although sunlight has a nasty way of damaging skin, you *need* sunlight. Your skin needs some sunlight regularly to keep your bones healthy. Yes, your bones. Obviously your bones can't soak up rays themselves. Bones need the skin to initiate the *vitamin D synthesis* process because bones need vitamin D in order to develop new bone cells properly.

Skin cells contain a molecule that's converted to vitamin D when ultraviolet rays strike the molecule. Then vitamin D leaves the skin and goes through the bloodstream to the liver and kidneys where vitamin D is converted to the hormone calcitriol. Going back into the bloodstream, calcitriol circulates through the entire body and regulates the amount of calcium and phosphorus present (keeping the levels of these important minerals within the normal range) — the reason that vitamin D is so important. Calcium and phosphorus are key to the development and maintenance of healthy bones. Without proper calcium and phosphorus metabolism, a condition called *rickets* can develop.

Rickets results in soft, curved bones that can't support the body's weight. Without vitamin D, calcitriol can't be produced, and without calcitriol, the calcium and phosphorus that's needed to build strong skeletons can't be metabolized. Rickets affects mostly children who don't eat properly or play outside. Fortunately, the milk that's now fortified with vitamin D helps to prevent this condition in children. However, a lack of vitamin D also affects adults.

In adults, problems with the liver or kidneys, where vitamin D is converted, can affect the production of calcitriol and result in a condition similar to rickets — *osteomalacia* — the softening of bones in adults. Other conditions that can lead to long-term vitamin D deficiency and osteomalacia include taking phenytoin (a seizure medication), genetic defects, and intestinal diseases that prevent vitamin D absorption. See Chapter 4 for more on pathophysiology of the bones.

## Healing your wounds

Say you cut yourself while chopping carrots. I've done this. It hurts. But, the pain doesn't last long, especially if you cut right through the nerve endings! However, the skin's wound must be repaired because bacteria can get into the cut, cause an infection, and compromise your health.

Luckily, your skin has great regenerative powers. Unlike starfish or earthworms, humans don't regenerate body parts, but some human organs are capable of growing new tissue. The liver and the skin's epidermis are capable of regenerating tissue. Wouldn't it be nice if the brain could regenerate some cells?

When you have a tiny, superficial surface wound (a little scratch), the epidermis simply replaces the damaged cells. In a few days, the scratch is gone. But, when you wound yourself deep enough that blood vessels are damaged, the healing process is a little bit more involved.

First, you bleed. The blood washes any debris or microbes right out of the wound and fills the wound area immediately. Just as quickly, the blood vessels around the wound constrict to slow down the blood flow so that too much blood doesn't seep out through the damaged vessels. A component of blood called *platelets* stick to the collagen fibers that make up the blood vessel wall and thus patch the blood vessel with a *platelet plug*.

After the platelet plug forms, a whole chain of events takes place to form a clot. Enzymes called *clotting factors* initiate the reactions that take place. There are 12 of these clotting factors, and the whole process is complex. Here's a rundown of what happens, focusing on the most important steps:

- **Prothrombin:** This clotting factor converts to thrombin. Calcium is required for this reaction.

- **Thrombin:** This factor acts as an enzyme and causes the plasma protein fibrinogen to form long threads called fibrin.

- **Fibrin threads:** Wrapping around the platelet plug, these threads form a meshlike template for a clot.

- **Clot:** The meshlike structure traps the red blood cells and forms a clot. As the red blood cells that are trapped on the outside of the clot dry out (or the air oxidizes the iron in them, like rust), they turn a brownish-red color and a scab forms.

Underneath the scab, the blood vessels regenerate and repair themselves, and in the dermis, cells called *fibroblasts* spur on the creation of new cells to regenerate the tissues in the damaged layers. Scars are created to provide extra strength to skin areas that were deeply wounded. Scar tissue has many interwoven collagen fibers, but no hair follicles, nails, or glands. Feeling is usually lost in the area covered with scar tissue because nerves are damaged.

## Controlling your thermostat

Your skin plays an important role in homeostasis (see Chapter 8), which is the process of adjustments your body goes through to keep everything in balance. Your body is continually converting energy from food (measured in calories) into energy in the form of ATP (see Chapter 3). But not all the energy from food is converted into ATP. About 60 percent of food energy is converted to heat. As the metabolic reactions that produce ATP occur inside each cell, heat is given off. This heat keeps your body temperature in the normal range as part of homeostasis.

Normal body temperature is 98.6 degrees Fahrenheit (37 degrees Celsius), but body temperature normally fluctuates between the range of 97 degrees Fahrenheit (F) and 100 degrees F. So don't get too upset if you have a temperature of 99.9 degrees F. Your skin performs two actions that help to increase your body temperature when you're cold and, likewise, decrease your body temperature when you're hot.

When your skin is cold, two things happen in the layers of your skin. First, the sweat glands are not activated, so heat is kept from escaping your body. Holding the heat inside, the skin raises your body temperature. Second, your blood vessels constrict so that your blood remains deep within the body and around vital organs. If the warm blood was flowing near the skin surface, heat could escape from the body. If you get really cold and your body temperature starts falling lower than 97 degrees F, you get goose bumps and start to shiver. (Refer to the "Enough to make your hair stand on end" sidebar earlier in this chapter.) Shivering, an involuntary contraction of muscles, helps you warm you up because muscles generate heat during contraction.

When you're hot, the blood vessels in your skin start to open so that excess heat can escape the bloodstream. Also, your sweat glands activate. The water in sweat absorbs heat from inside the body. While you sweat, the excess heat is removed from the inside of the body through the skin. The heat escapes the body as the sweat evaporates.

## Sensing what's going on

So how does your body know when it's cold or hot? How do you know when you get a cut or splinter? How can you tell the difference between being tickled with a feather and punched with a fist? The answer is that your skin contains specialized receptors for hot and cold, touch, pressure, and pain.

The skin's dermis contains nerve endings that serve as these specialized receptors (refer to Figure 6-1). Not every inch of skin contains all those types of nerve endings. So in some spots, you can sense the differences when you touch various objects; you can feel the difference between your skin being lightly pressed on or heavily pushed on; some spots sense cold, and some spots sense heat. All the receptors connect to a nerve that runs through the subcutaneous layer. The nerve relays information through the network of nerves up to your brain. You may yell "OW!" when you hurt yourself or elicit a giggle when you're tickled because your brain makes sense of the information.

# Skin Disorders and Diseases

Skin problems have a variety of causes. With the location of the skin on the body — that is, right out there in the open — the skin gets assaulted with germs that can cause infections. It also gets bombarded with the sun's rays, which can cause forms of skin cancer in some people. The skin also can be affected by genetic disorders or damaged by chemicals or fire. In this section, I give you some information on two of the most common but preventable skin problems: skin cancer and burns.

## Skin cancer

I know, I know. A few sections back, I said that you need sunlight to stay healthy. Now, I'm telling you about skin cancer. You ask, "What's the deal?" Well, you really do need sunlight to be healthy, but the line between healthy and unhealthy is a fine one. Just a few minutes of sunshine a day are all that's needed for the skin to make adequate amounts of vitamin D to keep your bones healthy. In the case of skin and sunlight, more isn't better.

With almost everything related to the body, moderation is key to good health, and sunscreen is a pretty good idea, too.

If you regularly get too much sun over many years, you have a good chance of developing skin cancer. Cell overgrowth is the mark of a disease known as *cancer* — that is, cells in an organ or in tissues grow too fast. In the skin, cancer is classified as a *melanoma* (a malignant or spreading type) or a *nonmelanoma* (limited to one portion of the skin) type.

> ✔ **Basal cell carcinoma:** (*Carcin-* is a root meaning cancer; a *carcinogen* is a cancer-causing substance.) The most common, this type of skin cancer occurs in the stratum germinativum (also called the *basal layer*), the lower part of the epidermis where skin cells originate. When ultraviolet radiation — whether it comes from the sun or a tanning booth — adversely affects the stratum germinativum, a tumor results. However, a tumor in the basal layer is sneaky because the body's immune system becomes increasingly unable to detect the tumor as it grows. This sneakiness allows the tumor to get bigger and bigger without the body's defenses trying to fight it. Luckily, after the tumor is noticed, it's usually easily cured. Signs of basal cell carcinoma include
>
> • Pale marks
>
> • Reddish patches that recur

- Round, smooth growths with a raised edge

- Shiny bumps

- Sores that don't heal

If you have a basal cell carcinoma removed, that doesn't mean it's gone for good. These carcinomas tend to recur. Fortunately, basal cell carcinoma is a nonmelanoma type of skin cancer, which means that although a cancerous tumor forms, it doesn't form in the melanocytes of the skin and is limited to just one part of the skin so there is a decreased chance of it spreading.

✔ **Squamous cell carcinoma:** The next most common, this type of skin cancer is a melanoma type that starts in the epidermal layer of the skin (where squamous cells are found). Squamous cell carcinoma is more likely to spread to a nearby organ than basal cell carcinoma, and squamous cell carcinoma carries a death rate of one percent. For every 100 people diagnosed with squamous cell carcinoma, only one is likely to die. The signs for this type of skin cancer are the same as those listed for basal cell carcinoma in addition to bleeding warts that develop a scab. This kind of skin cancer can recur, so avoidance of sunlight (to prevent further damage) and follow-up examinations (to check for early changes) are necessary for people diagnosed with squamous cell carcinoma.

✔ **Malignant melanoma:** The most dangerous, these skin cancer tumors start in the melanocytes (the cells that produce melanin), so they're heavily pigmented. An overgrowth of melanocytes cause normal circular, brown moles, but they don't spread. In contrast, malignant melanoma are so dark that they're almost black, and its borders aren't confined like a mole. These cancerous spots look like a spot on your garage floor where you spilled oil. Normal moles don't itch or hurt, but melanomas often do. Also, the skin around a melanoma often turns red, white, or gray.

As the popularity of sunbathing increased, so did the incidence of melanoma, which occurs most often in light-skinned people who have a history of severe sunburns, especially when they were kids. Compared to the 1 percent death rate of squamous cell carcinoma, 20 percent of people diagnosed with malignant melanoma die within five years.

To best protect yourself from the damaging rays of the sun, remember to

✔ Wear sunscreen with a sun protection factor (SPF) of at least 15.

✔ Wear a wide-brimmed hat (so your ears are protected as well as your face).

✔ Stay out of the midday sun (10 a.m. to 3 p.m.) to avoid the highest concentration of cancer-causing rays.

✔ Wear sunglasses that protect against both types of ultraviolet rays from the sun (that is, UV-A and UV-B). Not only can ultraviolet rays damage the eye itself, causing problems such as cataracts, the UV rays also wreak havoc on the very delicate skin around the eyes, causing hyperpigmentation (dark spots) and wrinkles.

# Burns

Heat, radiation, electricity, or chemicals can cause burns. If you ever shocked yourself on an electrical appliance or scrubbed your bathroom until the skin on your hands became red from the cleaner, you experienced a burn. Most people have been burned after touching something hot, whether the source of heat is an iron or a pot on the stove. The seriousness of a burn is measured by how many layers of skin (depth) the burn goes through and how much surface area the burn affects. After the determination of seriousness is made, classification as first-degree, second-degree, third-degree, or fourth-degree burns takes place.

✔ **First-degree burn:** Everyone is probably pretty familiar with sunburn — you're lucky if your experience isn't personal — but did you know that sunburn is actually a first-degree burn? A first-degree burn affects only the epidermis. This outer layer of skin reddens, and the pain receptors tell your brain of the damage, but blisters and swelling don't occur. These burns heal without scarring.

✔ **Second-degree burn:** Also called a *partial-thickness burn,* a second-degree burn affects the entire epidermis and the top part of the dermis. These burns cause pain and redness, and blisters occur. The more blisters there are, the deeper the burn is. Second-degree burns usually heal within 10 to 14 days and leave minimal scars if just a little bit of the dermis is affected, unless an infection occurs. If the burn goes deep into the dermis, healing can take anywhere from one to four months and scarring is probable.

✔ **Third-degree burn:** Also called a full-thickness burn, a third-degree burn goes through the entire epidermis and dermis and destroys the blood vessels, glands, hair follicles, and the pain receptors, so pain is nonexistent after the fact. The wound itself looks leathery and can range in color from white to tan, brown to black, or just plain red. Without blood vessels, healing is slow. Infection also occurs easily because the cells of the immune system that fight off the bacteria can't get to the tissue without the blood vessels transporting them. And, without sebaceous glands, the skin has no natural lubrication; thus, water loss is a problem. Also, without the hair follicles responding to changes in body temperature and helping to insulate the body, regulating heat loss becomes difficult.

✔ **Fourth-degree burn:** Leaving little chance for survival unless the damaged area is small, these burns not only go through all layers of the skin but also underlying muscle tissue or organs — all the way down to the bone.

To determine the amount of burn-affected body surface, physicians use the *rule of nines* (see Figure 6-2). The body's entire surface is divided up and measured in multiples of nine with everything except the genital region equaling 99 percent. The genitals amount to 1 percent of the body's surface. The entire head and neck region account for 9 percent (front = 4.5 percent, back = 4.5 percent). The front of the trunk accounts for 18 percent, and the back accounts for another 18 percent. The front of each arm accounts for 4.5 percent, and the back of each arm accounts for 4.5 percent with the total of arms, hands, and shoulders — front and back — equaling 18 percent. The front of the legs and feet equal 18 percent, and the back of the legs and feet equal another 18 percent. So when doctors say burns cover 45 percent of the body, you'll know how they arrive at their estimates.

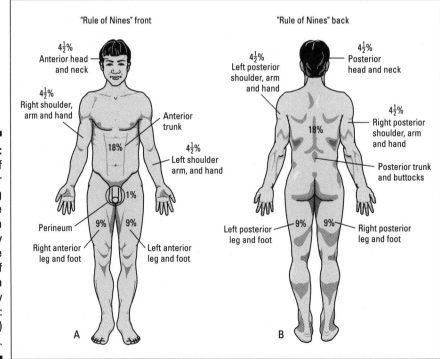

**Figure 6-2:** The Rule of Nines for determining percentage of area damaged by a burn. The value of each area of the body is shown: Front (A) and back (B).

*From LifeART®, Super Anatomy 2, © 2002, Lippincott Williams & Wilkins*

Treatment for burns involves prevention of complications while the patient heals. Giving the patient fluids intravenously prevents fluid loss. Keeping the patient warm prevents heat loss. Although it sounds odd, burn patients are usually put under heat lamps to keep them warm. Keeping the patient's visitors to a minimum and wrapping the burned areas in bandages soaked in antibacterial medications prevents bacterial infections.

When repairs need to be made, the damaged tissue needs to be sloughed off the patient and new skin is grafted onto the affected areas. Referred to as an *autograft* (*auto-* means self, as in autobiography), skin grafts can come from elsewhere on the patient's body or from another person, but rejection of the donated skin is then a problem. Donated skin is referred to as a *heterograft* (*hetero-* means other). More recently, however, skin has been grown in laboratories. The process is called *tissue culturing* because just a few epidermal cells are taken from the burn patient, and those cells are grown into sheets of epidermal tissue that are used for skin grafts to the patient. This method is considered autografting, but it provides much more tissue for serious burn victims than could be taken directly from their bodies. In less than one month, enough epidermis to cover the entire body can be grown from a piece of the patients' own tissue that is just three square centimeters in size. Amazing technology for the amazing anatomy.

# Part III
# Focusing on Physiology

The 5th Wave    By Rich Tennant

"Okay, you're really catching on to where all the body parts go, but we need to remember that there's quite a bit going on behind the scenes. Now, we're going to get good and gory and study the innards!"

# In this part . . .

*I*f the anatomy part of this book shows you how the framework, or chassis, of the body is put together, then this part about physiology covers the engine. The organ systems of your body work together to keep the entire body moving, just as a car engine keeps the entire car moving.

In Chapters 7 to 13, you take a look at each of the systems in a body: the nervous, endocrine, circulatory, respiratory, digestive, urinary, and immune systems, respectively. (I know, I left out the reproductive system, but check out Part IV for that.) Many bodily processes occur through cycles, and you can take a look at those cycles in these chapters.

Realizing that the same processes occur in the body repeatedly makes understanding anatomy and physiology much easier. When you're familiar with what is happening in your body and why, you may find that you have a feeling of control over your body and the choices that you make that affect it.

# Chapter 7

# Getting on Your Nerves: The Nervous System

*In This Chapter*

▶ Beginning with the brain and spinal cord: The CNS

▶ Networking nerves within a complex-but-organized system

▶ Relaying impulses through a cell and across a synapse

▶ Receiving impulses: Your five senses

▶ Degenerating tissues and neurofibrillary tangles

*T*hink about it. You couldn't do much without your nervous system. You couldn't run from a lion, tie your shoe, or rub your belly while patting your head. Your brain couldn't receive impulses (chemical messages) from your skin to let you know that you're cold or that you just touched something hot or sharp. You also couldn't feel the loving touch of your partner, mother, or child. Most important of all, you couldn't read this book!

But, thankfully, you can read this book, so your nervous system is functioning pretty darn well. This chapter tells you how your nerve network relays impulses to the different parts of the brain and controls some important processes as well as what happens when those functions go awry.

## Weaving a Well-Connected Web

Whenever I see a picture of the nervous system, a spider's web comes to mind. Take a look at Figure 7-1 and see what you think. Like the web that a spider spins, the nervous system is a well-organized group of connected strands and the nerves are thin but strong.

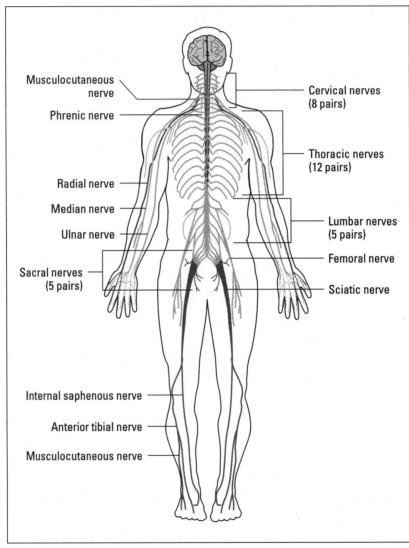

Musculocutaneous nerve

Phrenic nerve

Radial nerve

Median nerve

Ulnar nerve

Sacral nerves (5 pairs)

Internal saphenous nerve

Anterior tibial nerve

Musculocutaneous nerve

Cervical nerves (8 pairs)

Thoracic nerves (12 pairs)

Lumbar nerves (5 pairs)

Femoral nerve

Sciatic nerve

**Figure 7-1:**
The nervous
system.

*From LifeART®, Super Anatomy 1, © 2002, Lippincott Williams & Wilkins*

# Navigating the nervous system

REMEMBER

The nervous system is divided into the *central nervous system* (CNS), which consists of the brain and spinal cord, and the *peripheral nervous system* (PNS), which consists of all the nerves that project out from the brain and spinal cord. A *nerve* is a vessel that contains nerve fibers and connects them to the CNS. Nerve fibers can be of two types: *motor*, which send impulses away from the CNS, or *sensory*, which send impulses toward the CNS. Whereas the CNS is the brain and spinal cord, the PNS consists of

✔ **Cranial nerves,** which stem from the brain.

✔ **Spinal nerves,** which stem from the spinal cord.

✔ **Sensory fibers,** which are all over the body and send impulses to the CNS via the cranial nerves and spinal nerves.

✔ **Motor fibers,** which connect to muscles and glands and send impulses from the CNS via the cranial and spinal nerves.

The PNS is further divided into

✔ **The somatic system,** which consists of motor fibers sending impulses from the CNS to the voluntary skeletal muscles, as well as sensory fibers receiving input from receptors in the skin and initiating impulses as a reaction to the input (see information on receptors and effectors in the following section)

✔ **The autonomic system,** which consists of motor fibers sending impulses from the CNS to the glands, the heart, and involuntary smooth muscle (as in organs). The autonomic system is made up of the following:

- **The sympathetic nervous system:** Nerves originate in the thoracic and lumbar regions of the spinal cord (see Figure 7-2).

- **The parasympathetic nervous system:** Nerves originate in the brain and sacrum.

Both the sympathetic and the parasympathetic systems control internal organ functions that are involuntary and that happen subconsciously, such as breathing, heartbeat, and digestion.

## *Integrating the input with the output*

Before I describe the structure of a nerve and tell you how nerves send impulses (chemical messages), I want to tell you the major functions of the nervous system. The entire nervous system has just three jobs to do, and these jobs overlap.

### *Sensory input*

This function informs the CNS and helps organs adjust to what's going on inside and outside of the body. So if you see something that scares you, sensory input initiates the *fight-or-flight response*. Sensory input allows information from the body's receptors, such as in the skin (see Chapter 6), to create an impulse that shoots up to the spinal cord and then directly to the brain.

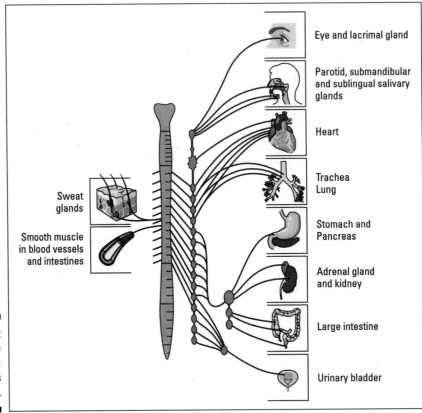

**Figure 7-2:**
The
sympathetic
nervous
system.

Say that you're walking down a dark alley late at night. (I don't know why you'd do that, but just pretend.) Suddenly, a big thug jumps out from behind a dumpster. You immediately sense danger because your eyes have received visual input of the external condition. Your brain shouts "Uh, oh!" and sends an impulse to your adrenal glands — via the sympathetic division of the autonomic part of the PNS — to release some adrenaline (see chapter 8) pronto. The adrenaline alters your internal condition, makes your heart beat faster, and gets blood flowing faster to your muscles in preparation for fighting or running away — the *flight* part.

## Integration

The CNS makes sense of the input it receives from all around the body. As *interneurons* (a neuron — nerve cell — situated between a sensory fiber and a motor fiber) integrate input from sensory nerves into the CNS, the CNS quickly assesses the situation and sends impulses out to nerves that can make an action happen.

### Motor output

The nervous system also stimulates muscles and glands to move and secrete substances, respectively. These action-making nerves — the ones that cause an effect to happen — are called *effectors*. Motor fibers are effectors in that they respond to information received by the sensory fibers.

## Beginning with a nerve cell

A nerve starts with a *neuron,* which is a nerve cell that moves the nerve impulses through the nervous system. *Neuroglial cells* (also called *neuroglia*), the other type of cell in the nervous system, plays a supporting role to neurons.

### Neurons

Figure 7-3 shows the structure of a motor neuron and a sensory neuron.

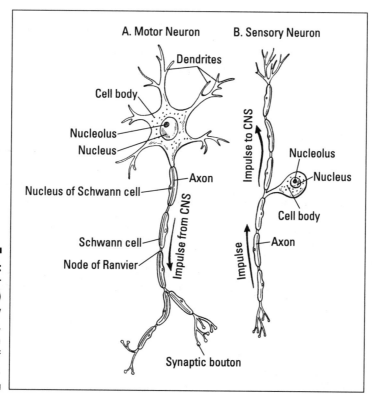

**Figure 7-3:**
Motor
neuron (A)
and sensory
neuron (B),
structure
and path of
impulses.

Neurons have three parts, although the look of a neuron varies with the specific function of the nerve it's in (that is, motor or sensory). The three parts of a neuron — any neuron — are the

- **Cell body,** which contains the nucleus and other organelles to keep a cell running (see Chapter 3).
- **Dendrites,** which receive information from other neurons and send impulses in the direction of the cell body.
- **Axons,** which are located on the opposite end of the neuron from the dendrites and send impulses away from the cell body.

### Neuroglial cells

The neuroglial cells provide the neurons with nutrients and help to protect the delicate threads of the nerve. Three types of cells that support the neurons:

- **Astrocytes** form connections between capillaries and neurons; thus neurons can get nutrients and get rid of wastes.
- **Microglial cells** flood injured sites to engulf microbes and remove waste.
- **Schwann cells** wrap around an axon like a tortilla wraps meat, beans, and cheese to make a burrito. Then they produce the myelin sheath that insulates and protects the nerves. You don't want your nerves to fray, do you?

# Transmitting Impulses

Nerve impulses have a domino effect. Each neuron receives an impulse and must pass it on to the next neuron and make sure the correct impulse continues on its path. Through a chain of chemical events, the dendrites pick up an impulse that's shuttled through the axon and transmitted to the next neuron. The entire impulse passes through a neuron in about seven milliseconds — faster than a lightning strike. Here's what happens in just six easy steps:

1. **Polarization of the neuron's membrane: Sodium is on the outside, and potassium is on the inside.**

   Cell membranes surround neurons just as any other cell in the body has a membrane. When a neuron is not stimulated — it's just sitting with no impulse to carry or transmit — its membrane is polarized. Not paralyzed. Polarized. Being polarized means that the electrical charge on the outside of the membrane is positive while the electrical charge on the inside of the membrane is negative. The outside of the cell contains excess sodium ions ($Na^+$); the inside of the cell contains excess potassium ions ($K^+$). (*Ions* are atoms of an element with a positive or negative charge. For more on ions, see Chapter 5.)

I know what you're thinking: How can the charge inside the cell be negative if the cell contains positive ions? Good question. The answer is that in addition to the $K^+$, negatively charged protein and nucleic acid molecules also inhabit the cell; therefore, the inside is negative as compared to the outside.

An even better question for you to come up with while reading this section is: If cell membranes allow ions to cross, how does the $Na^+$ stay outside and the $K^+$ stay inside? If this thought crossed your mind, you deserve a huge gold star! The answer is that the $Na^+$ and $K^+$ do, in fact, move back and forth across the membrane. However, Mother Nature thought of everything. There are *$Na^+/K^+$ pumps* on the membrane that pump the $Na^+$ back outside and the $K^+$ back inside. The charge of an ion inhibits membrane permeability (that is, makes it difficult for other things to cross the membrane).

2. **Resting potential gives the neuron a break.**

When the neuron is inactive and polarized, it's said to be at its resting potential. It remains this way until a stimulus comes along.

3. **Action potential: Sodium ions move inside the membrane.**

When a stimulus reaches a resting neuron, the gated ion channels on the resting neuron's membrane open suddenly and allow the $Na^+$ that was on the outside of the membrane to go rushing into the cell. As this happens, the neuron goes from being polarized to being *depolarized*.

Remember that when the neuron was polarized, the outside of the membrane was positive, and the inside of the membrane was negative. Well, after more positive ions go charging inside the membrane, the inside becomes positive, as well; polarization is removed and the threshold is reached.

Each neuron has a threshold level — the point at which there's no holding back. After the stimulus goes above the threshold level, more gated ion channels open and allow more $Na^+$ inside the cell. This causes complete depolarization of the neuron and an *action potential* is created. In this state, the neuron continues to open $Na^+$ channels all along the membrane. When this occurs, it's an *all-or-none phenomenon*. "All-or-none" means that if a stimulus doesn't exceed the threshold level and cause all the gates to open, no action potential results; however, after the threshold is crossed, there's no turning back: Complete depolarization occurs and the stimulus will be transmitted.

When an impulse travels down an axon covered by a myelin sheath, the impulse must move between the uninsulated gaps called *nodes of Ranvier* that exist between each Schwann cell.

4. **Repolarization: Potassium ions move outside, and sodium ions stay inside the membrane.**

After the inside of the cell becomes flooded with $Na^+$, the gated ion channels on the inside of the membrane open to allow the $K^+$ to move to the outside of the membrane. With $K^+$ moving to the outside, the membrane's repolarization restores electrical balance, although it's opposite of the initial polarized membrane that had $Na^+$ on the outside and $K^+$ on the inside. Just after the $K^+$ gates open, the $Na^+$ gates close; otherwise, the membrane couldn't repolarize.

5. **Hyperpolarization: More potassium ions are on the outside than there are sodium ions on the inside.**

   When the $K^+$ gates finally close, the neuron has slightly more $K^+$ on the outside than it has $Na^+$ on the inside. This causes the membrane potential to drop slightly lower than the resting potential, and the membrane is said to be hyperpolarized because it has a greater potential. (Because the membrane's potential is lower, it has more room to "grow."). This period doesn't last long, though (well, none of these steps take long!). After the impulse has traveled through the neuron, the action potential is over, and the cell membrane returns to normal (that is, the resting potential).

6. **Refractory period puts everything back to normal: Potassium returns inside, sodium returns outside.**

   The refractory period is when the $Na^+$ and $K^+$ are returned to their original sides: $Na^+$ on the outside and $K^+$ on the inside. While the neuron is busy returning everything to normal, it doesn't respond to any incoming stimuli. It's kind of like letting your answering machine pick up the phone call that makes your phone ring just as you walk in the door with your hands full. After the $Na^+/K^+$ pumps return the ions to their rightful side of the neuron's cell membrane, the neuron is back to its normal polarized state and stays in the resting potential until another impulse comes along.

Figure 7-4 shows the transmission of an impulse.

Like the gaps between the Schwann cells on an insulated axon, a gap called a *synapse* or *synaptic cleft* separates the axon of one neuron and the dendrites of the next neuron (refer to Figure 7-3). Neurons don't touch. The signal must traverse the synapse to continue on its path through the nervous system. Electrical conduction carries an impulse across synapses in the brain, but in other parts of the body, impulses are carried across synapses as the following chemical changes occur:

1. **Calcium gates open.**

   At the end of the axon from which the impulse is coming, the membrane depolarizes, gated ion channels open, and calcium ions ($Ca^{2+}$) are allowed to enter the cell.

2. **Releasing a neurotransmitter.**

   When the calcium ions rush in, a chemical called a *neurotransmitter* is released into the synapse.

### 3. The neurotransmitter binds with receptors on the neuron.

The chemical that serves as the neurotransmitter moves across the synapse and binds to proteins on the neuron membrane that's about to receive the impulse. The proteins serve as the receptors, and different proteins serve as receptors for different neurotransmitters — that is, neurotransmitters have specific receptors.

### 4. Excitation or inhibition of the membrane occurs.

Whether excitation or inhibition occurs depends on what chemical served as the neurotransmitter and the result that it had. For example, if the neurotransmitter causes the $Na^+$ channels to open, the neuron membrane becomes depolarized, and the impulse is carried through that neuron. If the $K^+$ channels open, the neuron membrane becomes hyperpolarized, and inhibition occurs. The impulse is stopped dead if an action potential cannot be generated.

If you're wondering what happens to the neurotransmitter after it binds to the receptor, you're really getting good at this anatomy and physiology stuff. Here's the story: After the neurotransmitter produces its effect, whether it's excitation or inhibition, the receptor releases it and the neurotransmitter goes back into the synapse. In the synapse, the cell "recycles" the degraded neurotransmitter. The chemicals go back into the membrane so that during the next impulse, when the synaptic vesicles bind to the membrane, the complete neurotransmitter can again be released.

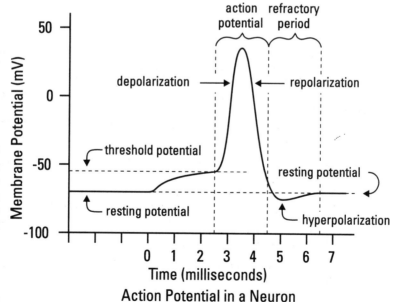

**Figure 7-4:**
Transmission of a nerve impulse: Resting potential and action potential.

Action Potential in a Neuron

# Thinking About Your Brain

I'm sure that you know where your brain is, but do you know what parts of your brain do what? Looking at a brain (see Figure 7-5), it all looks gray and soft, but the hard truth is that different areas of the brain are responsible for different functions.

The major parts of the brain are the *cerebrum, cerebellum, brain stem,* and *diencephalon.* The four connecting cavities in the brain are called *ventricles.* In this section, you can find out some details about the parts of your brain and its ventricles. Okay, Igor, let's go.

## Keeping conscious: Your cerebrum

If you're conscious, you're using your cerebrum. The cerebrum, the largest part of your brain, controls consciousness.

The cerebrum is divided into left and right halves, called the *right* or *left cerebral hemispheres,* and each half has four lobes: *frontal, parietal, temporal,* and *occipital.* The names come from the skull bones (see Chapter 4) that overlie these lobes. Table 7-1 shows you what each lobe controls.

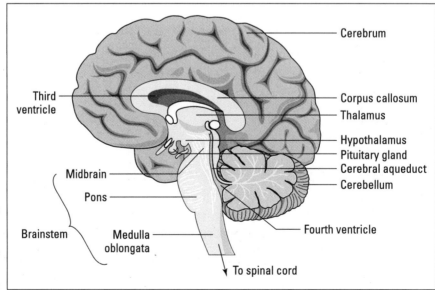

**Figure 7-5:**
Sagittal view of the brain.

*From LifeART®, Super Anatomy 1, © 2002, Lippincott Williams & Wilkins*

| Table 7-1 | Functions of Lobes within Cerebral Hemispheres | | |
|---|---|---|---|
| *Frontal Lobe* | *Parietal Lobe* | *Temporal Lobe* | *Occipital Lobe* |
| Speech production | General interpretation area | Interpretation of sensations | Recognizing objects visually |
| Concentration | Understanding speech | Remembering visually | Vision |
| Problem solving | Ability to use words | Remembering through sounds | Combining images received visually |
| Planning | Sensations felt on skin: heat/cold, pressure, touch, pain | Hearing | |
| Voluntary muscle control | Other sensations | Learning | |

The *cortex,* the brain's outer layer, covers the cerebrum and is gray because it contains cell bodies. The brain's curvy bumps are *gyri* (gyrus is singular). The shallow grooves that separate the bumps are called *sulci* (sulcus is singular). Deep grooves in the brain are *fissures.*

When you look at the top of a brain, you notice a deep groove running down the middle of the cerebrum. This groove is the *longitudinal fissure,* and it divides the cerebrum into the left and right hemispheres. The *corpus callosum,* located within the brain at the bottom of the longitudinal fissure, contains myelinated fibers that join the left and right hemispheres. Sometimes, people with severe epilepsy have their corpus callosum cut, so that the electrical misfirings that cause a seizure are limited to only half of their brain, giving them some normal functioning in the other half.

## Dancing the light fantastic: Your cerebellum

The cerebellum lies just below the bottom of the cerebrum. Like the cerebrum, a fissure divides the cerebellum into two parts and the outside is gray matter and the inside is white matter. (See "Gray and white matter matters" sidebar in this chapter.)

## Gray and white matter matters

The brain looks gray around its periphery because the cell bodies are located near the surface of the brain so that they can be near blood cells. These tissues of the brain and spinal cord are called *gray matter*. However, a protective sheath that makes nerves look white insulates axons, which send impulses. So they're referred to as *white matter*.

The cerebellum coordinates your muscle movements. If you can't dance, blame your cerebellum. Taking information from impulses received within the cerebrum, the cerebellum makes your muscle movements smooth and graceful. I'm not saying the cerebellum makes you glide across the floor like Ginger Rogers or Fred Astaire, but it keeps you from moving your limbs like a robot as you walk.

The cerebellum also maintains normal muscle tone. Not all your muscles can relax at one time, or you'd be a bag of bones and tissue. At all times — even when you are asleep — some of your muscles contract to help you maintain posture and muscle tone. The impulses that tell the muscles to contract are controlled by the cerebellum.

## *Coming up roses: Your brain stem*

Think of your brain as a beautiful rose — albeit a bumpy, slimy, gray one — and you can guess where your brain stem is. The brain stem consists of the *midbrain, pons,* and *medulla oblongata.* The medulla oblongata becomes the spinal cord after it passes through the hole called the *foramen magnum* (see Chapter 4) at the bottom of your skull.

Inside your brain, just in front of (anterior to) the cerebellum, lie the midbrain and pons. The midbrain serves as a "station" for information passing between the spinal cord and the cerebrum or between the cerebrum and cerebellum. Impulses pass through the midbrain, which has centers for reflexes based on vision, hearing, and touch. If you see, hear, or feel something that scares you, alarms you, or hurts you, the information is sent to your midbrain, which immediately responds by sending out impulses to generate the appropriate type of scream, jump, or exclamation. *Reflex arcs* sometimes create immediate, natural responses.

Reflex arcs (see Figure 7-6) happen automatically whenever you touch something really hot or too sharp. Sensory neurons detect pain, temperature, pressure, and the like. If sensory neurons detect something that could harm your body, such as heat that can cause a burn or a sharp object that could puncture the skin, an impulse passes from the receptor in the skin through the sensory neuron to the spinal cord and then to motor neurons that cause a muscle to contract and pull the body part at risk of injury away from the heat or sharp object.

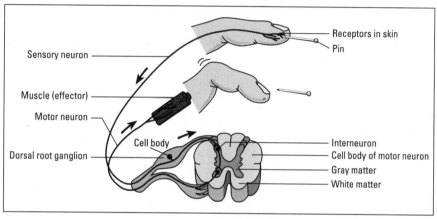

**Figure 7-6:**
A reflex arc.

*From LifeART®, Super Anatomy 3, © 2002, Lippincott Williams & Wilkins*

Reflexes occur so fast that you don't have time to think about how to react. But the other reason for why you don't even think about it is that the impulse doesn't make it to your brain in time to generate a reaction. By the time the impulse gets to your brain, the spinal cord has already taken care of the problem! In normal processes of the CNS, impulses travel to the brain for interpretation and production of the proper response. However, using the spinal cord rather than the brain to produce a response, reflex arcs save time and possibly damaging consequences.

If the midbrain is a station for impulses, the pons is the bridge that joins the cerebellum with the cerebrum and the left and right hemispheres of the cerebrum. Axon bundles fill the pons (not the ponds!) Those axons respond quickly to information received through the eyes and ears.

The medulla oblongata, which becomes the spinal cord, is responsible for several important functions, such as the beating of your heart, your breathing, and the regulation of your blood pressure. The medulla oblongata — don't you just love that name? — also contains the axons that send out the signals for coughing, vomiting, sneezing, and swallowing, based on information it receives from the respiratory or digestive systems. And, whenever you get those annoying hiccups, blame your medulla oblongata.

## Trickling through the ventricles

A *lateral ventricle* is on each side of your brain. The other two ventricles are, believe it or not, the third and fourth ventricles. The *third ventricle* lies just about in the center of your brain; the *fourth ventricle* lies at the top of the brain stem. The *cerebral aqueduct* (also called the aqueduct of Sylvius) connects the third and fourth ventricles. The cerebral aqueduct becomes the central canal of the spinal cord.

When you think of aqueduct, you may think of Rome. The Roman aqueducts were built as a system for distributing water. Well, in your CNS, the ventricles and aqueducts serve as a system to circulate *cerebrospinal fluid* (CSF).

A clear fluid made in the brain, CSF is contained in the four ventricles of the brain, the subarachnoid space, and the central canal of the spinal cord. The CSF serves to pick up waste products from the CNS cells and delivers them to capillaries so that the bloodstream can dispose of the wastes. The CSF also cushions the CNS. Along with your skull and vertebrae, the CSF adds a protective layer around your brain and spinal cord.

Perhaps the most important function of CSF is to keep the ions in balance and thus stabilize membrane potentials. CSF circulates from the lateral ventricles of the brain to the third ventricle, through the cerebral aqueduct into the fourth ventricle, and then down through the central canal of the spinal cord. From the fourth ventricle, CSF oozes into the subarachnoid space just under the arachnoid membrane, which continuously covers the spinal cord and brain. In the subarachnoid space in the brain, the CSF can seep through tiny spaces to get to the bloodstream.

Commonly known as a *spinal tap,* CSF is drawn through a needle for analysis from the subarachnoid space. The CSF can be tested for the presence of bacteria that cause meningitis or for the presence of proteins that can indicate other diseases, such as Alzheimer's.

## Regulating systems: The diencephalon

Right smack in the middle of the brain, the *hypothalamus* and the *thalamus* that lie in the walls and floor of the third ventricle form the diencephalon. The hypothalamus regulates sleep, hunger, thirst, body temperature, blood pressure, and fluid level, and maintains homeostasis, which keeps the body's systems in check and in balance. Think of the hypothalamus as the computerized sensor in your car that makes your warning lights come on if the water in your radiator is too low, the engine overheats, or the oil is low. The hypothalamus also controls when the pituitary gland signals the endocrine system to secrete hormones (see Chapter 8).

## The most interesting system of all

If you ever fell in love, enjoyed sex, stashed away some great memories, or felt enraged (seems like the full cycle of a relationship, doesn't it?), then your limbic system was in use. The limbic system is not an anatomic structure, but rather a collection of areas in the brain — certain parts of the cerebrum and diencephalon — that are involved in some emotional issues. These areas control your interest in sex (libido), memory, pleasure or pain, and feelings such as happiness or sorrow, fear, affection, and rage. Although those reactions and emotions may not be key to your survival, they do make life interesting.

The thalamus is the gateway to the cerebrum. Whenever an impulse travels from somewhere in your body (except from the nose; sensations of smell are sent directly to the brain by the olfactory nerve), it passes through the thalamus. The thalamus then relays the impulse to the proper location in the cerebral cortex (refer to Table 7-1), which then interprets the message. Think of the thalamus like an e-mail server, routing your message through the correct lines.

# Making Sense of Your Senses

The brain controls the five senses — touch, hearing, sight, smell, and taste. The skin — one of the sense organs — contains receptors that act as sensors that transmit information regarding touch up through the nervous system to the brain for processing and reaction. Likewise, each of the other sense organs — ears, eyes, tongue, and nose — contain specialized types of receptors (Table 7-2). Chapter 6 is devoted to the skin, so this section outlines how the remaining four sense organs work.

| Table 7-2 | Receptors Found in Sense Organs | |
|---|---|---|
| **Sense Organ** | **Function** | **Receptor** |
| Ears | Mechanoreceptors | Cilia in the ears detect movement of the eardrum and ossicles (ear bones) that allow you to hear. |
| Eyes | Photoreceptors | The retina of the eye detects light to allow vision. |

*(continued)*

**Table 7-2** *(continued)*

| Sense Organ | Function | Receptor |
|---|---|---|
| Tongue | Chemoreceptors | Taste buds detect various chemical molecules present in foods. |
| Nose | Chemoreceptors | Cilia in the nasal cavity detect chemical molecules present in the air. |
| Skin (touch, pressure, pain, temperature) | Mechanoreceptors (touch and pressure), thermoreceptors (hot or cold), and nociceptors (pain) | Specialized nerve endings detect different sensations (see Chapter 6 for more on skin) |

# Hearing

Your ears (see Figure 7-7) allow you to hear sounds because of their structural design. Your outer ear acts as a funnel to channel sound waves to the eardrum. When sound waves cause the eardrum to vibrate, the ear bones, called *ossicles,* receive and amplify the vibration and then transmit the vibration to the inner ear. The vibrations create tiny ripples in the fluid of the inner ear, and the movement makes the *cilia* (very teeny, tiny hair cells) bend. Because the hair cells move, they're referred to as *mechanoreceptors.*

As the hair cells bend, they move a thin membrane called the *tectorial membrane.* The movement of the tectorial membrane generates impulses to be sent from the nerve fibers of the inner ear to the *cochlear branch* of the *vestibulocochlear nerve* (one of the cranial nerves), which connects directly to the brain stem. The brain stem then passes the information — technically referred to as *auditory stimuli* — on to the area of the cerebrum that interprets the impulse as a specific sound. This entire process happens nearly instantly; little delay lies between the sound wave's initiation and your brain's interpretation of the stimulus as a sound.

Your ears also transmit information to your brain as to what position you're in; that is, whether you're horizontal or vertical, spinning or still, moving forward or backward. Therefore, your ears are the key organ of *balance.* The process of transmitting information to the brain about your body position is basically the same as that for hearing. When you're moving, the fluid of the inner ear moves and causes hair cells to bend. The bending of the hair cells initiates an impulse being sent from the nerve fibers of the inner ear. Your brain then processes the information about where you are spatially and initiates movements to help you keep your balance.

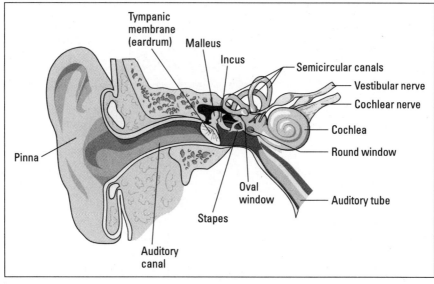

**Figure 7-7:**
Anatomy
of the ear.

## Seeing

Vision is probably the most complex of the senses. Your eye's *pupil* (see Figure 7-8), the dot in the center of your eye that's usually black in color, must allow the correct amount of light in. The *iris,* the pretty, colored part of your eye, contains the muscle that controls pupil dilation. The iris contracts to dilate the pupil and allow more light in, such as when you're in a darkened room or outside at night. Likewise, another muscle in the iris contracts to make the pupil smaller to let less light in, such as when you're in the sun or a brightly lit place. The *cornea* covers the iris and the pupil. (You can kind of see a clear area if you look at an eyeball from the side.) The eye *lens* — the dark area that you see when you look into someone's pupil — is behind the pupil.

A clear, gelatinous material fills the *vitreous body,* which lies behind the lens. And I'm not joking when I tell you that the gelatinous material is *vitreous humor.* The transparent vitreous humor gives the eyeball its rounded shape, and it also lets light pass through it to the back of the eyeball. The *retina* resides at the back of the eyeball, and the retina is what contains the two types of photoreceptors — *rods,* which detect dim light and are sensitive to motion, and *cones,* which detect color and fine details. Three types of cones detect color; one each for detecting red, blue, and green. A missing or damaged cone (regardless of the type) results in color blindness.

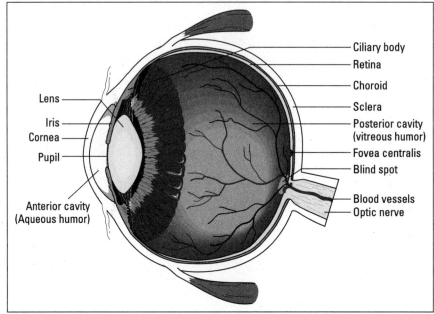

**Figure 7-8:**
Internal
structures
of the eye.

Ciliary body
Retina
Choroid
Sclera
Posterior cavity
(vitreous humor)
Fovea centralis
Blind spot
Blood vessels
Optic nerve

Lens
Iris
Cornea
Pupil
Anterior cavity
(Aqueous humor)

When light strikes the rods and cones, nerve impulses are generated and sent to cells that form the optic nerve. The optic nerve joins your eyeball directly to your brain and sends impulses to the brain for interpretation in the occipital lobe.

## Smelling

The nose knows, but the *olfactory cells* know better. The nose is the sense organ for smell (*olfaction* is the proper term), but only the olfactory cells that line the top of your nasal cavity detect foreign odors in the air you breathe — foreign odors from dirty diapers or from food that's cooking on the stove or from baking in the oven or any other substance with an odor. As you take in air through your nostrils, those odorous gases waft right up to your olfactory cells where the chemicals bind to the cilia that line your nasal cavity. That action initiates a nerve impulse being sent through the olfactory cell, into the *olfactory nerve fiber,* up to the *olfactory bulb,* and right to your brain. (The olfactory bulb is the expanded area at the end of the olfactory tract where the olfactory nerve fibers enter the brain.) The brain then "knows" what the chemical odors are from, and you'll know what you're smelling.

### Eating carrots: Mother was right

Eating carrots really improves your ability to see at night.

When the rods inside the retina detect movement and dim light, they release a chemical called *rhodopsin* that splits into the protein *opsin* and the pigment *retinal*. Opsin and retinal generate nerve impulses that are sent to the brain. Seeing in the dark is hard at first because your eyes are adjusting to the change. After making more rhodopsin so that impulses can be sent to the brain, you can see. So the more rhodopsin you have, the quicker you adjust to darkness.

What's the connection to carrots? Retinal, one of the byproducts of rhodopsin that generates the nerve impulses, is derived from vitamin A, and carrots are loaded with vitamin A.

## Tasting

The tongue is the sense organ for taste, but taste and smell usually go hand in hand. If you can't smell (like when you have a cold), and you don't see, your brain doesn't have any information to interpret. The sense of taste works similarly to the sense of smell. The tongue's *taste buds* — which actually lie in the grooves between the bumps on your tongue — contain chemoreceptors for sweet, sour, bitter, and salty. Whenever you put food or drink into your mouth, the taste buds can detect what's in it. When the elements in food bind to the *microvilli* (tiny projectile fibers) in a taste bud, a nerve impulse is generated and carried through the sensory nerve fiber at the end of the microvilli. The impulse makes its way to the brain, of course. The brain then interprets the impulse as a certain food and causes the release of the digestive enzymes needed to break down that food. So the sense of taste is tied to the endocrine system as well as to the digestive system and serves a higher function than just allowing you to enjoy fine dining.

# Pathophysiology of the Nervous System

Much can go wrong with your nervous system. Brain tumors or severed spinal cords certainly cause problems, but subtle diseases and disorders that progress slowly and imperceptibly often plague parts of the nervous system. This section sheds light on a few of those for you.

## Multiple sclerosis

Multiple sclerosis (MS) affects the myelin sheath that covers the axon of a nerve. The myelin sheath develops a lesion that becomes inflamed and irritated. When the lesion heals, hard scar tissue — called a sclerosis — is left behind. As the disease affects more and more nerves, the number of scleroses increases, leading to multiple damage sites.

The hard scar tissue interferes with the nerve's ability to conduct an impulse through the axon. So if an impulse can't be transmitted, a movement or response can't occur. As the disease progresses, movement becomes increasingly difficult and then impossible.

Most people with MS begin to see signs in young adulthood to middle age. It takes years or decades for the disease to progress to its most serious stage. Fortunately, several prescription medications are now available to treat MS.

## Macular degeneration

Macular degeneration is a vision disorder that's now a leading cause of blindness. It's becoming more and more common because this disorder usually strikes older individuals. As an increasing percentage of the population enters their elderly years, the incidence of macular degeneration also increases. What happens to people with this disorder is that the *macula lutea* — a small area of the retina with a large concentration of cones (photoreceptors that detect color and fine details) — weakens and degenerates. One of the telltale signs of macular degeneration is when straight objects, like a tree trunk or lamppost, appear wavy. Objects look smaller or larger than they really are, and colors fade. Normally, the macula lutea allows for sharp vision and detection of bright colors, but older people with macular degeneration have blurry vision.

One cause of macular degeneration is overgrowth of new blood vessels around the macula lutea. New growth sounds healthy, but these vessels aren't. They leak, and as they ooze blood, more of the macula lutea is destroyed. Another cause is excessive sun exposure; people with lighter pigment in their eyes (blue or green) are affected more severely. Smoking and high blood pressure can also lead to leaky blood vessels. Laser treatment may stop the overgrowth of the leaky blood vessels for a while and zinc supplements are thought to keep macular degeneration from getting worse after a diagnosis is made, but no overall, permanent cure is really available.

## Alzheimer's disease

Everybody forgets once in a while. Where you put your keys, the name of someone from high school, or the date and time of an appointment may all slip your mind from time to time, but can you imagine your mind slipping away? That's what Alzheimer's disease is like. It's hard to believe that people could ever forget their children or closest friends, but Alzheimer's disease makes that happen. Eventually, Alzheimer's patients can't care for themselves because they forget how to perform simple tasks and become mentally disturbed.

People with Alzheimer's disease tend to have bundles of a fibrous protein tangled around the nucleus of their neuron cell bodies. These *neurofibrillary tangles* are thought to be made of a certain protein — dubbed *Alzheimer disease-associated protein* (ADAP) — that can often be detected in the CSF of these patients. *Amyloid plaques* — globs of protein that surround the axon branches — are another sign of Alzheimer's disease.

As more and more neurofibrillary tangles and amyloid plaques affect the brain's neurons, the symptoms of Alzheimer's disease worsen. The ability to remember people, places, events, and objects, as well as the ability to reason come from the frontal lobe and the limbic system — the areas of the brain that show the abnormal neurons during an autopsy of an Alzheimer's patient. A cure or way to prevent Alzheimer's disease isn't yet available.

# Chapter 8

# Moaning About Hormones: The Endocrine System

*N*ot all hormones make women crabby and emotional, and women aren't the only ones who have hormones. They aren't *her*-mones after all. Men and children have hormones, too. This chapter explains what a hormone is, what hormones do, where they come from, and how they get to where they need to go. Then at the end of the chapter, you can find out what happens when hormones go awry, Okay? I won't forget to put that in there! Why do you always think I'll forget something? Don't you trust me? You must not love me anymore! Oh, sorry. It's just my hormones acting up again.

## Honing In on Hormones

Okay, I promised I'd tell you what a hormone is. A *hormone* is a chemical substance created by an organ or tissue of the endocrine system (usually an endocrine gland) that, after being secreted from an endocrine gland, is carried by the bloodstream to other cells, where the hormone brings about its effect. The *endocrine system* is the body's system of glands that produces hormones (see Figure 8-1). Throughout this chapter, you can see that the nervous system (see Chapter 7) and the endocrine system work closely together to regulate your body's systems. The nervous system controls when the endocrine system should release or withhold hormones, and the hormones control the metabolic activities within the body. This section guides you through the types of hormones, how hormones work, and the role hormones play in homeostasis.

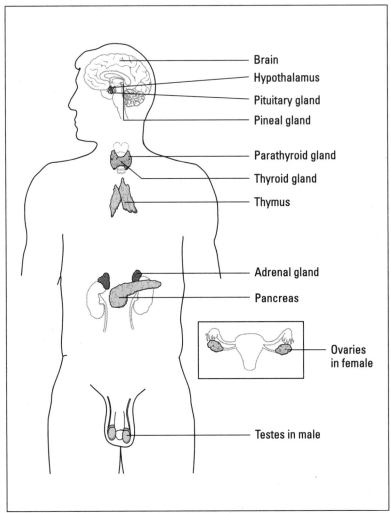

Brain
Hypothalamus
Pituitary gland
Pineal gland

Parathyroid gland
Thyroid gland
Thymus

Adrenal gland
Pancreas

Ovaries
in female

Testes in male

**Figure 8-1:**
The
endocrine
system

*From LifeART®, Super Anatomy 1, © 2002, Lippincott Williams & Wilkins*

## Looking at types of hormones

Glands in the endocrine system produce two major classes of hormones:

✔ **Steroid hormones** are lipids created from cholesterol.

✔ **Nonsteroid hormones** are made of amino acids and proteins. (See Chapter 2 for more information on amino acids and proteins.)

Steroid hormones, such as estrogen, progesterone, testosterone, aldosterone, and cortisol, are the hormones that people most often think of as "raging" through the system or causing PMS (premenstrual syndrome). Nonsteroid hormones can be divided into the following four different types:

✔ **Protein hormones:** Insulin, growth hormone, and prolactin

✔ **Amine hormones:** Epinephrine and norepinephrine

✔ **Glycoprotein hormones:** Follicle-stimulating hormone (FSH), luteinizing hormone (LH), and thyroid-stimulating hormone (TSH).

✔ **Peptide hormones:** Antidiuretic hormone (ADH) and oxytocin

Table 8-1 discusses the source and function of the most important steroid and nonsteroid hormones.

| Table 8-1 | Important Hormones: Source, Functions | |
|---|---|---|
| *Hormone* | *Source* | *Function(s)* |
| Adrenocorticotropic hormone (ACTH) | Pituitary gland (anterior part) | Stimulates growth of cortex in the adrenal gland and secretion of corticosteroids by the cortex of the adrenal gland; increased during stressful periods |
| Antidiuretic hormone (ADH) | Pituitary gland (posterior part) | Stimulates the kidneys to reabsorb water, preventing dehydration |
| Calcitonin | Thyroid gland | Targets the bones, kidneys, and intestines to reduce the level of calcium in the blood |
| Corticotropin (see ACTH above) | | |
| Epinephrine/ norepinephrine | Medulla of the adrenal gland | Stimulates the heart muscle and other muscles during the *fight-or-flight* response; increases the amount of glucose in the blood (to provide quick energy to fight or flee) |
| Estrogen/progesterone | Ovaries | In addition to stimulating production and release of eggs, these hormones target muscles, bones, and skin to develop female secondary sex characteristics |

*(continued)*

**Table 8-1** *(continued)*

| Hormone | Source | Function(s) |
| --- | --- | --- |
| Glucagon | Pancreas | Causes liver, muscles, and adipose tissue to release glucose into the bloodstream |
| Glucocorticoids | Cortex of the adrenal glands | These steroid hormones (such as cortisol) target all tissues to release glucose and raise the blood glucose level when necessary |
| Growth hormone (GH), also called somatotropin | Pituitary gland (anterior part) | Targets the bones and soft tissues to promote cell division, synthesis of proteins, and growth of bone tissue |
| Insulin | Pancreas | Causes liver, muscles, and adipose tissue to store glucose as a way of lowering blood glucose level |
| Melatonin | Pineal gland | Targets a variety of tissues to mediate control of biorhythms, the body's daily routine |
| Mineralocorticoids (such as aldosterone) | Cortex of the adrenal glands | These steroid hormones target the kidney cells to reabsorb sodium and excrete potassium to keep electrolytes (ions) within normal level |
| Oxytocin | Pituitary gland (posterior part) | Present just before and after giving birth; stimulates the uterus to contract and mammary glands (in breasts) to release milk |
| Parathyroid hormone | Parathyroid glands | Stimulates the cells in bones, kidneys, and intestines to release calcium so that blood level of calcium increases |
| Prolactin | Pituitary gland (anterior part) | Targets the mammary gland to stimulate production and secretion of milk |
| Testosterone | Testes | In testes, it stimulates the production of sperm; in skin, muscles, and bones, it causes development of male sex characteristics |

| Hormone | Source | Function(s) |
| --- | --- | --- |
| Thyroid-stimulating hormone | Pituitary gland (anterior part) | Stimulates the thyroid gland to pro duce and release its important hormones, calcitonin and thyroxin |
| Thyroxin | Thyroid gland | Distributed to all tissues to increase metabolic rate; involved in regulation of development and growth. |

# Discovering how hormones work

When a gland secretes a hormone, the hormone gets shuttled through the bloodstream to its *target cell,* which is where the hormone has its effect. The hormone controls the activity that goes on in the target cell. Like the relationship between a resident dormitory director and a student, the hormone controls the activity that goes on in the target cell. The director provides the student with information and rules. If everything is quiet, calm, and moving along according to the rules, the director lets the student continue without interference. If the student has a loud party, causes a scene or problem, or breaks the rules, however, the director makes sure the student's activity is curtailed.

The target cell is the student, and the hormone is the dormitory director. The hormone comes along regularly to provide "information," but if the student has too much or not enough going on in that dorm room, the hormone provides a different type of information to reduce or increase the activity.

Of course, the way that hormones stimulate target cells is a bit more complicated than a dormitory director knocking on your dorm room door. Steroid hormones activate cells in a different manner than nonsteroid hormones. To understand how both hormone types stimulate target cells, look at the list that follows:

- ✔ **Steroid hormones** pass through a cell membrane and bind with receptor molecules (the molecule that receives the hormone) in the cell's cytoplasm. The receptor molecules with hormones attached move together into the cell's nucleus, where they activate certain genes to synthesize the needed protein. (See Chapter 2 for more on genes and protein synthesis.)

- ✔ **Nonsteroid hormones** have a bit harder time. Composed of amino acids and proteins, these hormones must join with a receptor on the target cell's membrane. Then the hormone — the first messenger — is taken into the cell via active transport. (See Chapter 3 for more information on active transport.) ATP, the energy molecule, is expended to carry the hormone inside. When the hormone gets inside, the production of a compound called *cyclic adenosine monophosphate* (cyclic AMP) — the second messenger — is stimulated, thus causing the target cell to produce the necessary enzymes.

When the desired effect has been achieved, the endocrine gland has to be "told" to stop producing the hormone. Otherwise, the opposite effect may occur.

For example, if your blood glucose level is too high, your pancreas secretes the hormone insulin, which removes some of the glucose from your bloodstream. But after the glucose level is back within normal range, the pancreas has to stop producing insulin or your glucose level may drop to below normal. Having too much of a substance can be dangerous, but having too little of a substance can be equally dangerous. So *negative feedback mechanisms* shut down the secretion of hormones after they've done their job.

The negative feedback mechanism works much like the thermostat in your home. If you set your thermostat to 65 degrees, the heat goes on when the temperature in your house drops below 65 degrees. The heat runs until the temperature warms up to 65 degrees again, and then the heat pump shuts off The thermostat shuts off because of feedback received from the thermometer. It doesn't keep running until you shut it off, otherwise the temperature in your home could climb to 90 degrees.

Your brain's hypothalamus detects your blood glucose level and sends feedback to the pancreas. (See the following "Changing to stay normal: Homeostasis" section.) When the hypothalamus senses that your blood glucose level is within normal range, it informs the pancreas to stop producing insulin.

## *Changing to stay normal: Homeostasis*

*Homeostasis* is the all-important set of checks and balances that your body goes through to stay healthy. Hormones play a big part in homeostasis. Hormones are transported around the body through the bloodstream. When the blood passes certain checkpoints in the nervous system (such as the hypothalamus inside the brain), hormone levels are "measured." It's like putting the dipstick into your car's engine to measure the oil.

If the level of a certain hormone is too low, the gland that produces that hormone is stimulated to produce the hormone. If a hormone level is too high, the gland that produces the hormone is "told" to stop production. The hormone comes from the endocrine system, but the stimulation and messages to change what's going on come from the nervous system. And don't forget the importance of the circulatory system! Without bloodstream transportation, none of the detection and implementation of changes would be possible.

Your body is always keeping tabs on the metabolic processes going on inside it. If anything (such as body temperature, glucose level, pH level) leaves the normal range, the checkpoints involved in homeostasis work with the endocrine system to bring the body's systems back into balance.

# Grouping the Glands of the Endocrine System

The body's glands are *exocrine glands* or *endocrine glands* as follows:

- **Exocrine glands** produce substances, but not hormones, that have an effect in the same tissues where they're produced.

  For example, the sebaceous gland is an exocrine gland. The sebaceous gland produces and secretes oils that keep hair and skin soft so they won't dry out. The oils don't travel around the body and don't have an effect anywhere else except the hair follicle where they're produced. The oils can be secreted quickly, but their effect doesn't last long.

- **Endocrine glands** produce hormones that are secreted into the bloodstream and travel to the place where they exert their effect.

  However, endocrine glands not only produce hormones, they must distribute them around the body. And what travels around the body continuously? Your blood. The bloodstream is like a monorail system that encircles an amusement park or zoo. Each blood cell is like a monorail car that picks up passengers in one location and drops them off in another. It might take a little while to get the hormone to the target cell, but the effect it has lasts longer than that of a substance secreted from an exocrine gland.

  The body has several endocrine glands, and each gland is discussed in the upcoming sections of this chapter. Some glands are recognizable as organs, but note that organs can act as glands and secrete hormones, too. The following list includes the endocrine glands from head to toe. (Well, make that head to pelvis — your legs and toes have no endocrine glands.)

  - Hypothalamus and pituitary glands
  - Thyroid gland
  - Thymus

- Pancreas
- Stomach
- Intestines
- Adrenal glands
- Ovaries and testes

## The task masters: The hypothalamus and pituitary glands

The hypothalamus is often called the "master gland" because it ultimately controls homeostasis. However, its hormones don't target the cells in the body that make the adjustments to achieve balance. Instead, the hypothalamus detects the level of substances in the blood and secretes hormones into the pituitary gland. The pituitary gland then releases the hormones that affect homeostasis. Figure 8-2 shows the relationship between the hypothalamus and pituitary glands.

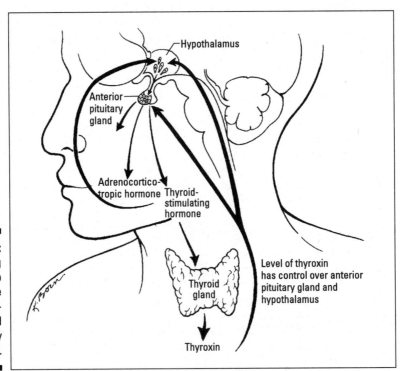

**Figure 8-2:**
The working relationship of the hypotha-lamus and pituitary gland.

The hypothalamus and the pituitary glands adjoin in the central portion of the brain called the *third ventricle* (see Chapter 7). The pituitary gland (also called the *hypophysis*) has two parts that secrete different hormones. The *anterior portion* of the pituitary gland is sometimes called the *adenohypophysis*. The *posterior portion* of the pituitary gland is the part that's directly connected to the hypothalamus (refer to Figure 8-2) and is sometimes referred to as the *neurohypophysis*.

The hypothalamus connects the nervous system with the endocrine system. The hormones that the posterior pituitary gland releases are actually made in the nerve cell bodies of the hypothalamus. The hormones travel down the axons of neurons that end in the posterior pituitary.

One of the hormones made in the hypothalamus but released by the pituitary gland is antidiuretic hormone (ADH), also called *vasopressin*. Homeostasis tries to ensure that the blood contains the proper amount of water. The hypothalamus contains special cells that act as sensors to detect when the blood has too little water. When the blood needs more fluid, the hypothalamus produces ADH, and ADH travels down the axons into the posterior pituitary gland, which releases ADH into the bloodstream. ADH travels through the bloodstream until it reaches its target — the kidney cells.

Via active transport, ADH goes into the cells of the kidney's tubules (little tubes). Inside those cells, it alters the reactions so that more water is removed from the urine being produced in the kidney, allowing the water to be absorbed into the bloodstream. Removing water from the urine makes the urine more concentrated. This also can cause problems, but less serious problems than having too little water in your bloodstream. Not drinking enough water, though, can lead to chronic dehydration, which leads to chronically concentrated urine, and then problems (such as kidney stones, ouch!) may appear. Give your hypothalamus and pituitary gland a break — be sure to drink the recommended half-gallon of water every day to help keep your fluid level within the normal range.

Small blood vessels connect the anterior portion of the pituitary gland to the hypothalamus, which releases the hormones *hypothalamic releasing hormone* and *hypothalamic release-inhibiting hormone*. When the hypothalamus sends hypothalamic releasing hormones down into the anterior part of the pituitary gland, the pituitary gland is stimulated to secrete one of its hormones. The anterior pituitary secretes some pretty important hormones, such as growth hormone, prolactin, and TSH (refer to Table 8-1).

The anterior pituitary gland secretes some other hormones, including melanocyte-stimulating hormone (MSH); gonadotropic hormones, such as follicle-stimulating hormone (FSH) and luteinizing hormone (LH); and adrenocorticotropic hormone (ACTH). MSH targets the melanocytes in the skin and

## Diabetes insipidus: Not from sipping too much

*Diabetes* is a disease that results in the glucose level of the blood being outside the normal range. Most people are familiar with *diabetes mellitus,* which is caused by a problem with insulin that results in the glucose level being too high. *Diabetes insipidus* is a similar problem with a different cause. Diabetes insipidus is caused by the inability of the hypothalamus to produce the proper amount of antidiuretic hormone (ADH), which is responsible for stimulating the kidney to return water to the bloodstream. Water is removed from the urine being produced in the kidney, and then the water is absorbed into the bloodstream. Without ADH, however, very little water is removed from the urine and returned to the bloodstream. This makes the concentration of substances (such as glucose) much higher in the bloodstream, and it results in large amounts of watery urine that carries electrolytes right out of the body. Despite the large amounts of urine they produce, people with diabetes insipidus really are dehydrated and thirsty. They can be treated, however, with ADH.

stimulates them to produce melanin, the skin pigment. The gonadotropic hormones target the gonads — the sex organs — and stimulates them to produce the male and female sex hormones and gametes. ACTH targets the cortex of the adrenal glands and stimulates it to secrete its hormones. The anterior pituitary releases the gonadotropic hormones and ACTH, but their function is to stimulate the release of other hormones from other glands in the body. The same is true of growth hormone and TSH — they're only intermediaries to the release of other hormones.

When the anterior pituitary gland's hormones have stimulated other glands to produce their effects, the hypothalamus secretes hypothalamic release-inhibiting hormone into the anterior pituitary. This hormone makes the anterior pituitary gland stop stimulating the other glands.

## Thy thyroid and thou

Your thyroid gland looks somewhat like a butterfly that straddles your trachea (windpipe). Each lobe of the thyroid gland — the butterfly's wings — is to the side of the trachea; a stretch of tissue called the *isthmus* joins both lobes. Columnar epithelial cells (see Chapter 3) line the secreting part of the lobes; these cells secrete a jellylike substance called *thyroglobulin.*

Thyroglobulin serves as a mixing medium. As blood carrying iodine (from the food, such as table salt, that you eat) passes through the thyroid gland (at a rate of four to five liters per hour!), the thyroglobulin "traps" the iodine in the bloodstream. Then the iodine interacts with the amino acid tyrosine to form *thyroxin* and *triiodothyronine,* hormones secreted by the thyroid gland.

Thyroxin is not secreted as often as triiodothyronine, but when it is, its action lasts longer. And much more thyroxin is present in the thyroid gland than triiodothyronine. When TSH (thyroid-stimulating hormone) urges the thyroid hormones to get out there and do their jobs, the hormones adsorb (not *ab*sorb; adsorb means "stick to") protein molecules in the blood and are released slowly. Thyroxin has many important jobs to do; here are a few of thyroxin's responsibilities:

- Regulating the rate at which cells metabolize and respire (use oxygen and release carbon dioxide)
- Increasing the rate at which the cells use glucose and stimulating the conversion of the storage form of glucose (that is, glycogen) into glucose so that the blood level of glucose increases
- Helping to maintain body temperature (The more reactions that are going on in your body, the higher your body temperature.)
- Playing a role in growth and differentiation of tissues in children and teens
- Increasing the amount of certain enzymes in the mitochondria that are involved in oxidative reactions
- Influencing the metabolic rate of proteins, fats, carbohydrates, vitamins, minerals, and water
- Stimulating mental processes

So as you can see from the above list of important functions, thyroxin and thyroid-stimulating hormone are pretty important to your overall well-being. In the "Pathophysiology of the Endocrine System" section, I show you what can happen if you have too much or too little of the thyroid hormones.

## *Thy must know the thymus*

Like the thyroid gland, the thymus is a lobed gland. It's situated in the thoracic cavity just below your collar bones and just above your heart. The main function of the thymus gland is that it turns an immature T lymphocyte from the bone marrow into a fully mature T cell (see Chapter 13). The thymus produces a group of hormones, called *thymosins,* involved in differentiating and stimulating cells of the immune system. Recent AIDS research has focused on possible uses of thymosins.

## *Flat as a pancreas*

Your pancreas is a fibrous, elongated, flat (as a pancake) organ that lies nestled in your abdomen near your kidneys, stomach, and small intestine. Figure 8-1, presented earlier in this chapter, identifies its location. It's made up of two tissue types, and it has two different functions:

✔ **Digestive function:** Produce digestive enzymes that it then secretes into the small intestine (see Chapter 11).

✔ **Endocrine function:** Produce the hormones *insulin* and *glucagon,* which it secretes directly into the bloodstream.

Insulin and glucagon are involved in maintaining the glucose level of the blood. Insulin helps to decrease the glucose level; glucagon serves to raise it.

When you ingest food, your body immediately starts breaking down the molecules in the food to their smallest components. Glucose, a simple sugar, is the smallest component of a carbohydrate. It's what your body runs on; it's the brain's fuel and cells throughout your body convert glucose into the energy molecule ATP. (The details of digestion are in Chapter 11.) So glucose is pretty important to your ability to function. However, the amount of glucose in your blood has to be within a certain range; if there's too much or too little, your health is compromised.

Immediately after you finish eating something, insulin secretions deal with the glucose influx. Insulin works in four ways to help keep your glucose level within the normal range:

✔ Stimulates your cells to absorb glucose and metabolize it. This keeps it from floating free in the bloodstream.

✔ Stimulates liver and muscle cells to store glucose in the form of glycogen, thus removing even more glucose from the bloodstream.

✔ Stimulates fat and protein production, which requires glucose, thereby using up some more that is tubing down the bloodstream.

✔ Inhibits the use of fats and proteins for energy so that glucose remains stored for the future; glucose is locked away safely out of the bloodstream.

When a problem exists with the production or release of insulin, the glucose level of the blood can get pretty high. This situation — called *hyperglycemia* — means that there is a problem with homeostasis, and a disease results. The disease is diabetes, and I explain it for you in the "Pathophysiology of the Endocrine System" section later in this chapter.

Insulin is secreted just after a meal to shuttle away the glucose, but between meals, your glucose level may drop a bit below normal. You know the signs: Your stomach starts growling, you get a little shaky, and a headache starts. To manage this situation — called *hypoglycemia* — your pancreas releases some glucagons, which has the opposite effect of insulin; it puts stored glucose back into the bloodstream to raise the blood glucose level back to the normal range. Where does it get the stored glucose? All the cells and tissues where insulin shuttled the glucose — your liver, muscle cells, and fatty tissue.

 The body's ability to keep itself in balance is simply amazing. To help keep yourself in balance, don't eat huge meals that overwhelm your pancreas to produce more insulin and don't snack between meals. Give your glucagon a chance to do its job!

## Stomaching another gland

You may not think of the stomach as a gland, but it does secrete hormones. This makes the stomach one of those organs that double as a gland. Yes, the stomach is a key organ in digestion, but it also secretes hormones used during digestion, making it a gland. The stomach secretes a group of hormones called *gastrin*. There are many types of gastrin molecules — small, medium, and large, so to speak. Gastrin is responsible for stimulating the secretion of gastric acid, the main juice that mixes with the food in the stomach during digestion (see Chapter 11). Gastrin also causes the sphincter muscle — a muscle that surrounds an opening — at the bottom of the esophagus to contract, thereby controlling when food can be dumped into the stomach. And gastrin causes growth of the mucosal cells that line the stomach and secrete acid.

## Testing the intestines

So are the intestines an organ or a gland? This is another trick question because the answer is both: The intestines are an organ of the digestive system, but they also secrete hormones. The small intestine and the hypothalamus secrete *cholecystokinin* (CCK), which targets the gallbladder and stimulates the release of bile. Bile is a greenish-brown liquid that mixes with the food substances being digested to help reduce acidity. (The acids of the stomach are too harsh for the intestines, so bile, which contains bicarbonate, helps to neutralize the acids.) Bile also dissolves cholesterol in food, and makes fats easier to absorb. CCK also stimulates secretion of pancreatic enzymes.

It's no secret: The small intestine secretes another hormone called *secretin.* Secretin works similarly to CCK in that it stimulates the cells of the pancreas to secrete neutralizing substances. The intestines release enzymes that can't work in an acidic environment, so the actions of CCK and secretin ensure that digestion can continue.

# Topping the kidneys: Your adrenal glands

The root *renal* means kidney; the prefix *ad* means near. So guess where the adrenal glands are? Yep, near the kidneys. More specifically, they sit right on top of the kidneys (refer to Figure 8-1). Like your two kidneys — a left and a right — there are two adrenal glands, one on each side of the body.

Like the skin on a peanut or kidney bean, a thin capsule covers the entire adrenal gland. Inside, the adrenal glands have two parts: the *cortex* and the *medulla,* both of which have different functions. The functions of the cortex and medulla differ because they developed differently. When you were a wee little embryo, your adrenal cortex developed from the same tissue as your kidney and connective tissues (muscle, bone, cartilage, and so on). But your adrenal medulla developed from the same tissue as part of your autonomic nervous system (see Chapter 7).

### Adrenal cortex

This part of your adrenal gland secretes *corticosteroids,* which include *mineralocorticoids* and *glucocorticoids.* One of the most important mineralocorticoids is *aldosterone,* which is responsible for regulating the concentration of electrolytes, such as potassium ($K^+$), sodium ($Na^+$), and chloride ($Cl^-$) ions. This regulation keeps the salt and mineral content of the blood in check — more homeostasis!

*Electrolytes* are substances that split apart into *ions* (atoms of an element with a positive or negative charge) when in solutions (such as the watery tissue fluid around your cells or the cytoplasm within your cells). Electrolytes are capable of conducting electricity; hence, their name.

Aldosterone targets the kidneys' tubules and stimulates the reabsorption of $Na^+$. When the sodium ions are reabsorbed into the bloodstream, chloride ions quickly follow. ($Na^+$ and $Cl^-$ love to be together as NaCl, commonly known as "salt.") And where the salt is, water follows. If salt ions move into the bloodstream, water does, too, and the fluid level of the blood is increased. These actions control the body's fluid balance. Ultimately, the fluid and electrolyte balance affects blood pressure (see Chapter 9).

*Cortisol,* the main glucocorticoid hormone, is responsible for regulating the metabolism of proteins, fats, and carbohydrates. Most often, cortisol is released when you're stressed emotionally, physically, or environmentally. It affects metabolism in the following ways:

✔ Breaking down protein, decreasing protein synthesis (see Chapter 2), and moving amino acids from tissues to liver cells

✔ Moving fat from adipose tissue to bloodstream or increasing the rate of fat storage in adipose tissue

✔ Reducing the rate at which cells use glucose, which causes the liver to store more glucose as glycogen

By decreasing the number of immune cells circulating in the blood and decreasing the size of the lymphoid tissue, cortisol (and other corticosteroids) affect the immune system. Under severe stress and with large amounts of glucocorticoids circulating in the blood, the lymphoid tissue is unable to produce antibodies (see Chapter 13). Perhaps this explains why you always seem to get a cold when you're under pressure?

### Adrenal medulla

This part of the adrenal gland developed from the same tissues that turned into part of the sympathetic nervous system. So it should come as no surprise that some of the functions of the adrenal medulla involve regulating actions of the structures containing these nerves. The adrenal medulla produces a class of hormones called the *catecholamines,* of which *epinephrine* and *norepinephrine* are the major hormones.

Epinephrine, also called *adrenaline,* is known for initiating the *fight-or-flight response.* I'm sure you've experienced a rush of adrenaline at one time or another, usually when you're frightened. If something makes you jump or feel out of control, you can almost feel the adrenaline coursing through your bloodstream.

Without using glucose, epinephrine causes your heart rate and breathing rate to increase. Glucose is reserved for the brain's use. Epinephrine stimulates the release of free fatty acid molecules from your adipose tissue. Your muscles then use the free fatty acids for energy. Again, glucose is saved for your brain. After all, you need to think if you're in a scary situation.

Like epinephrine, norepinephrine is a catecholamine that's tied to the nervous system. Norepinephrine causes vasoconstriction, that is, tightening of the blood vessels. When the blood vessels get smaller, blood pressure increases. Therefore, norepinephrine is released when your blood pressure is low (hypotension), as well as when you're stressed.

## Going for the gonads

Your gonads — *ovaries* if you're female or *testes* if you're male — are the lowest endocrine glands in your body (refer to Figure 8-1). The ovaries produce and secrete *estrogen* and *progesterone;* the testes produce and secrete *testosterone*. Collectively, estrogen, progesterone, and testosterone are known as the *sex hormones*. Testosterone is also called an *androgen* because it stimulates male characteristics.

You may think that estrogen is limited to female animals, but you'd be wrong. Estrogen can be detected in the urine of male animals, and even, surprisingly, in growing plants! I hope that Venus flytraps don't get PMS and become man-eating plants!

Estrogen starts being secreted in females at a higher rate during puberty. It's responsible for initiating the development of the secondary sex characteristics of the female — structures that take a girl into womanhood. (Primary sex characteristics would be the presence of a vagina, uterus, and ovaries, which are obvious at birth.) The breasts enlarge, and fat is deposited in them; estrogen stimulates this process. Bone tissue grows rapidly, and height increases. Estrogen helps this process; estrogen causes calcium and phosphate to be retained in the bloodstream (so they can be used for bone growth) and stimulates the activity of the osteoblasts (new bone cells; see Chapter 4). Estrogen also allows the pelvic bones to widen. Yes, this means the hips get wider, but it also allows passage of a fetus during childbirth. (See "Why women have bigger hips than men" sidebar in Chapter 4.) In addition, estrogen increases the deposition of fat around the body, thus giving women a more rounded appearance than men. Gee, wider hips and more fat; that's not fair!

By causing changes in uterine secretions and in storing nutrients in the lining of the uterus, progesterone works to prepare the uterus for implantation of a fertilized egg. By promoting the growth of the lobules inside the breast and causing the cells to grow and become ready to secrete milk, progesterone also contributes to breast development.

Testosterone causes the development of secondary sex characteristics in males; primary sex characteristics are the penis and testicles at birth. But after a boy hits puberty, his muscle tissues start to grow, his sex organs enlarge, and hair develops on his chest and face and the hair on his arms and legs becomes darker and coarser. Both women and men develop hair under their arms and in their pubic area. Just as men also have some estrogen, women also have some testosterone. For more on development during puberty, turn to Chapter 15.

# *Pathophysiology of the Endocrine System*

Homeostasis is important to your health, and the nervous system and endocrine system work together to maintain homeostasis. Sometimes, though, homeostasis breaks down. A homeostasis malfunction is a *disease*. Some diseases, known as *local diseases,* only affect certain body parts. But other diseases, known as *systemic diseases,* affect an entire organ system or the entire body. If a disease develops and passes quickly, it's called an *acute illness.* If a disease develops over a long period of time and causes ill effects for a long time, it's a *chronic disease.* In this section, I give you information on some diseases that affect the endocrine system or result from malfunctions in the endocrine system.

## Diabetes mellitus

*Diabetes* comes from the Greek words meaning "siphon," "through," and "to go," all of which refer to the fact that people with diabetes tend to urinate excessively. *Polyuria,* the technical term for excessive urination, is a symptom in the two most common forms of diabetes: *diabetes insipidus,* which is a disorder of the pituitary gland, and *diabetes mellitus,* which stems from a pancreatic disorder. (See "Diabetes insipidus: Not from sipping too much" sidebar in this chapter.) Diabetes mellitus is the most common form of diabetes; so common in fact, that when you see the word "diabetes" alone, it's referring to diabetes mellitus (DM).

DM can be either of two types; the difference is in whether insulin is required for survival and why. Type 1 DM is the insulin-dependent form; type 2 DM is the non-insulin-dependent form. Genes, poor diet, obesity, pancreatic disease, pregnancy, hormonal problems, some drugs (such as steroids and oral contraceptives), and possibly even infection can cause diabetes.

Insulin is the hormone that shuttles glucose into the cells around the body, contributes to protein synthesis, and stimulates the storage of free fatty acids in adipose tissue. If the pancreas doesn't produce enough insulin or the body doesn't respond well to its own insulin, the metabolism of carbohydrates, protein, and fats is disturbed.

In a person with type 1 DM, the pancreas produces little or no insulin, and the body's tissues literally starve. One of the first symptoms of type 1 DM is unexplained loss of muscle tissue and fat from under the skin. Signs of type 2 DM develop more gradually.

Abdominal obesity is common in people with type 2 DM, as are factors including family history of diabetes, giving birth to infants weighing more than 9 pounds, stress or trauma, viral infection, and a history of diabetes when pregnant.

Both type 1 and type 2 DM patients experience excessive thirst and hunger, weight loss, fatigue, weakness, vision problems, itchy skin, and skin infections that don't heal well.

Dehydration is a symptom of diabetes (both type 1 and type 2) because the lack of insulin leading to excessive glucose in the bloodstream causes a high level of substances in the blood *(hyperosmolarity)*. And the way that the body deals with hyperosmolarity is to remove more water with the hopes of removing the excessive substances. Unfortunately, the water loss further contributes to the dehydration, and the problem is compounded.

Increased urine production stems from the breaking down of proteins in an attempt to get glucose for energy. (Remember that without insulin, cells utilize glucose poorly.) As the proteins are metabolized, the amino acids split apart and convert to glucose and urea in the liver. The extra urea leads to extra urine being produced.

Treatment for diabetes consists of insulin, a carefully monitored diet, and an exercise plan for patients with type 1 DM. Patients with type 2 DM often begin with a carefully monitored and regular eating plan, as well as regular exercise. Eating meals with roughly the same amount of carbohydrates at consistent times every day keeps the insulin level fairly even. By burning off excess glucose, exercise helps diabetics to lose weight (obesity is a contributing factor) and increase sensitivity to the insulin they already make. If diet and exercise fail to do the trick for type 2 diabetics, oral medications can help to control the glucose and insulin levels.

Complications of diabetes include retinopathy that can lead to blindness, heart problems, high blood pressure, and damage to the nerves of the feet. Because pain can't be felt as easily in the foot after the nerves are damaged, a person with diabetes is at risk for severe skin infections that can develop gangrene and require amputation.

## Hypothyroidism versus Graves' disease

Hypothyroidism and Graves' disease are two opposite problems. *Hypothyroidism* is a condition that results from the thyroid gland functioning below normal; Graves' disease results from *hyperthyroidism* — the thyroid gland functioning at an above-normal level. Graves' disease results in an enlarged thyroid gland, a goiter, and swelling of the eye muscles, thus the eyeballs protrude from their sockets. Hyperthyroidism makes a person irritable, nervous, and unable to sleep. Treatment options for patients with hyperthyroidism include oral medications, a single dose of radioactive iodine, or surgery to reduce the size and activity of the thyroid gland.

Patients with hypothyroidism may have a defect in their thyroid gland *(primary hypothyroidism)*, or their hypothalamus or pituitary glands may not be sending the proper releasing or stimulating hormones to cause secretion of the thyroid hormones *(secondary hypothyroidism)*. People with primary hypothyroidism may have inflammatory conditions similar to arthritis or chronic conditions, such as *Hashimoto's thyroiditis* (a disease in which the body's immune system attacks cells of the thyroid gland). Dietary deficiency of iodine and medications that negatively affect the thyroid gland also may cause secondary hypothyroidism.

Hypothyroidism has many signs and symptoms because the thyroid hormones have such a widespread effect. Nearly every cell in the body is stimulated by the hormone thyroxin, which regulates the rate of metabolism. And metabolism alone involves many reactions; thyroxin is necessary for most of these to take place at the proper levels. Symptoms of hypothyroidism are given in Table 8-2.

| Table 8-2 | Symptoms of Hypothyroidism Throughout Course of Disease | |
|---|---|---|
| *Initially* | *As Disease Progresses* | *Severe* |
| Fatigue | Decreased sex drive | Psychiatric problems, changes in behavior |
| Cold sensitivity | Stiff joints | Carpal tunnel syndrome |
| Weight gain without increase in food intake or decrease in exercise | Muscle cramps | High cholesterol, poor circu lation, heart problems |
| Constipation | Weight loss | Dry skin and hair, some hair loss; brittle, grooved nails |
| Memory problems | Numbness or tingling sensations | Impaired fertility |
| | | Weak colon, intestinal obstruction, anemia |

*Myxedema* is the most troublesome complication of hypothyroidism. Patients can slip into a coma because their breathing rate and heart rate slow down so severely. Because they're exchanging carbon dioxide and oxygen at a lower rate, the amount of carbon dioxide in their blood rises. This condition requires emergency intervention and is often fatal.

Treatment for people with hypothyroidism involves lifelong administration of a synthetic thyroid hormone. However, therapy must begin gradually so that the heart isn't negatively affected.

# Chapter 9

# Putting Your Heart in It: The Circulatory System

*W*hat do the Autobahn, Interstate 95, Route 66, Amtrak, and your circulatory system have in common? They're all transport pathways. The transport vehicles — cars, trucks, trains, and blood — certainly differ, but someone or something goes from one place to another. Whether it's people in a car, an 18-wheeler full of groceries, or blood cells carrying oxygen, cargo is moved along a path. This chapter tells you nothing about interstates or delivery trucks, but it does tell you what makes up your blood, what your blood and heart do, and how your blood picks up and delivers materials around your body.

## Carrying Cargo: Your Blood and What's in It

People may say that you have acting, music, dance, writing, or some other passion "in your blood," but do you know what's *really* in your blood? Tiny cells, tinier particles, and plenty of water. When a test tube of blood is *centrifuged* (spun really fast until the materials separate), the formed elements lie on the bottom of the tube, and the plasma floats on top. In this section, I take you through those layers to find out what's really in there.

COOL BODY BITS

## The hematocrit

Whenever you have a blood test done, one of the characteristics of your blood that the laboratory personnel check for is your *hematocrit* (HCT). The HCT is the percentage of your centrifuged blood that contains the formed elements, primarily the red blood cells. Normally, your HCT should be about 45 percent. If your HCT is low, you have too few formed elements in your blood. If your HCT is higher than normal, you have too many formed elements in your blood.

Blood — that deep maroon, body-temperature-warm, and coppery-tasting liquid that courses through your body — is a vitally important life-supporting, life-giving, life-saving substance that everybody needs. And every adult-size body contains about five quarts of the precious stuff. But blood is an anomaly. Because it consists of different types of cells based in a matrix — in blood, the matrix is *plasma* — blood technically is a connective tissue (see Chapter 3). However, because blood is a fluid, not many people think of blood as a tissue. In fact, the only fluid part of blood is the plasma; the different types of cells — *red cells, white cells,* and *platelets* — are referred to as *formed elements*. Think of the plasma as a river and the cells and platelets as inner tubes floating downstream. A peaceful picture, isn't it?

## *Watering down your blood: Plasma*

REMEMBER

About 92 percent of your plasma is water. The remaining 8 percent or so is made up of *plasma proteins,* salt molecules, oxygen and carbon dioxide gases, nutrients (glucose, fats, amino acids) from the foods you take in, urea (a waste product that becomes a major part of urine; see Chapter 12), and other substances being carried in the bloodstream, such as hormones or vitamins.

The three types of *plasma proteins,* created in the liver, have some important functions:

- **Albumin:** The smallest plasma protein and the most abundant, albumin maintains the proper osmotic pressure in the bloodstream. Having the proper osmotic pressure in the bloodstream ensures that a good balance between solutes (particles) and solution (liquid), and that helps to maintain the proper pH (a measure of acidity or alkalinity based on the concentration of hydrogen ions in a solution; see Chapter 12). Osmosis is explained in Chapter 3; briefly, osmosis is the action of water following solutes. The albumin, along with certain electrolytes floating in the plasma (such as sodium and potassium ions), act as solutes, and water moves from the body's tissues into the blood.

✔ **Fibrinogen:** During the process of clot formation, fibrinogen is converted into threads of fibrin, which then form a meshlike structure that traps blood cells to form a clot.

✔ **Immunoglobulin:** This is another word for antibody (see Chapter 13). Immunoglobulin proteins are created in response to an invading microbe, whether it's a virus or bacteria. When your body encounters an invader for the first time, immunoglobulins are formed as your immune system fights off the microbe. These immunoglobulins continue to circulate in the blood so that the next time you encounter that same microbe, you can attack quickly and not get sick.

## *Transporting oxygen: Red cells*

If you've ever seen red blood cells up close via microscope, you may remember that they actually do look like little inner tubes — this makes it easy to visualize them floating through your bloodstream. A *red blood cell* (RBC) has a technical name — *erythrocyte* — but it doesn't have one feature that most cells have: nuclei. RBCs contain *hemoglobin,* which allows them to do their very important job — carry oxygen. Hemoglobin is a protein formed from a globin molecule and four molecules of heme, which contains iron. The iron in the heme binds oxygen so that hemoglobin can carry it through the blood.

When oxygen binds to hemoglobin, the compound *oxyhemoglobin* forms. Oxyhemoglobin is bright red and provides the color you see in blood coming from arteries. When oxyhemoglobin releases oxygen, the remaining compound is *deoxyhemoglobin.* This substance has a deep purple color. This purple color is why your veins look blue. (Check out your wrists.)

RBCs don't live long. Each RBC, which is made in your red bone marrow, has about a four-month life span. When an RBC has lived its four months and is at its end, a *phagocyte* (a large cell that "eats" cellular debris and carries it to where it can be removed from the body) engulfs it. The phagocyte carries the dead RBC to the liver or spleen, where the RBC is destroyed. The liver or spleen destroys 2.5 million RBCs per second; fortunately, the red bone marrow produces as many just as fast. As the cell is being degraded, the hemoglobin is released and breaks into the proteins *globin* and *heme.* The iron is extracted from the heme and is recycled back to the red bone marrow to be put into newly developing RBCs.

## Staying well: White cells

REMEMBER

Although red blood cells get the credit for carrying oxygen and making blood red, white blood cells (WBCs), also known as *leukocytes,* do their fair share to keep you healthy. Your body doesn't have as many WBCs as RBCs, but the white ones are usually bigger, and it's a good thing; WBCs are the cells that have to fight off the bullies in your bloodstream — the invading microbes, such as bacteria and viruses (see Chapter 13).

Several types of WBCs are *basophils, eosinophils,* and *neutrophils,* which are *granular leukocytes* because they have tiny protein particles that prominently surround their nuclei. (Yes, WBCs have nuclei.) *Lymphocytes* and *monocytes* are *agranular leukocytes;* although these WBCs do have the protein particles around their nuclei, they aren't as obvious. The protein particles contain enzymes and other substances that act as natural antibiotics that help to kill bacteria and viruses.

Because WBCs function as part of the immune system (see Chapter 13), they not only exist in the blood but also in the tissue fluid that surrounds cells and lymph (which is excess tissue fluid that's absorbed by vessels of the lymphatic system; see Chapter 13). So when you have an infection in your skin, some WBCs can squeeze out of your tiniest blood vessels (capillaries) and get right to the site of infection. Table 9-1 tells you the specific functions of each type of WBC.

| Table 9-1 | White Blood Cell Functions |
|---|---|
| *Type of White Blood Cell* | *Specific Function(s)* |
| *Granular leukocytes* | |
| Eosinophils | Numbers increase during allergic reactions and parasitic infections |
| Neutrophils | First to respond to infections; they phago-cytize (eat and discard) bacteria and cellular debris |
| Basophils | Seep out of blood vessels into tissues at site of injury; release histamine to dilate blood vessels in area (allowing more oxygen, nutrients, and immune cells to get to injured tissues and speed repair) |

| Type of White Blood Cell | Specific Function(s) |
|---|---|
| *Agranular leukocytes* | ———— |
| Lymphocytes | ———— |
| B lymphocytes | B lymphocytes form antibodies used to fight infections (see Chapter 13) |
| T lymphocytes | T cells help to keep you healthy by destroying cells that contain foreign material, referred to as antigens (see Chapter 13) |
| Monocytes | Largest of the white blood cells (WBCs); mature into macrophages, cells of the immune system that engulf disease-causing microbes and spur other WBCs into action |

## Plugging along with platelets

Platelets are tiny pieces of cells. Large cells that exist in the red bone marrow — *megakaryocytes* — break into fragments, which are the platelets. Their job is to begin the clotting process and plug up injured blood vessels. Each platelet, also called a *thrombocyte,* has a short life span; they live only about ten days.

## Squeezing blood from a bone

Blood cells are made in your red bone marrow. The process that forms blood cells is *hematopoiesis,* and it goes on in the red marrow of your vertebrae, ribs, and skull, as well as at the ends of your long bones (such as your femur). *Stem cells* are in the red marrow, and those original stem cells divide and change (differentiate) until they become specific blood cells.

Red blood cells are formed like this:

1. **A stem cell (called *multipotent* because it has the potential to become many different types of cells) in the red marrow divides into two *myeloid stem cells.***

2. **A myeloid stem cell can go on to become a B or T lymphocyte (see Table 9-1) or it can differentiate into an *erythroblast, megakaryoblast,* or *myeloblast.***

   An *erythroblast* matures into an erythrocyte, an RBC.

   A *megakaryoblast* matures into a megakaryocyte, which fragments into platelets.

   A *myeloblast* can differentiate into any one of four types of white blood cells: basophils, eosinophils, neutrophils, or monocytes.

3. **After blood cells have matured in the bones, they enter the circulatory system, which then transports them around the body.**

# Figuring Out Cardiac Anatomy: Your Heart

The circulatory system — or *cardiovascular system* — consists of the heart and the blood vessels. The heart, the main organ of the circulatory system, causes blood to flow. The heart's pumping action squeezes blood out of the heart, and the pressure it generates forces the blood through the blood vessels. Anatomically speaking, the heart is only about the size of your fist. And, it's not really shaped like a heart; a human heart is really shaped like a cone (see Figure 9-1). It lies between your lungs, just behind (posterior to) your sternum, and the "tip" (apex) of the cone points to the left. In fact, your heart is situated slightly to the left of center in your chest.

A thick layer of muscle tissue and a protective membrane that folds into two layers, called the *pericardium* or *pericardial membranes,* surround the heart. The heart itself is a well-organized grouping of hollow spaces.

- ✔ **Endocardium:** This innermost layer of the heart is made up of *endothelial tissue* that lines the inside of the heart and is continuous with all your blood vessels.

- ✔ **Pericardial cavity:** Working your way out, in this space, you find the coronary vessels — the blood vessels that supply the tissues of the heart with nutrients and oxygen.

- ✔ **Myocardium:** This next layer out is the hard-working, contracting, muscular layer of your heart (*myo-* refers to muscle).

- ✔ **Epicardium:** This inner layer of the pericardium covers the myocardium. The epicardium secretes *pericardial fluid,* which protects the tissues as they rub together when the heart beats.

✓ **Parietal pericardium:** Beyond the pericardial cavity, working your way out to the outside of the heart, this outermost layer of the heart is a thin, white covering made of fibrous connective tissue that joins the major blood vessels (such as the aorta) to the sternum and diaphragm. Your heart is not just floating in your chest. Like the epicardium, the *parietal pericardium* also secretes *pericardial fluid* to lubricate the heart's tissues.

The hollow spaces of the heart are called *chambers*. Four chambers, two on each side, fill with blood and release blood in a rhythmic pattern:

✓ Left atrium

✓ Right atrium

✓ Left ventricle

✓ Right ventricle

Together, the left atrium and the right atrium are the *atria* (plural). A membrane called the *interatrial septum* separates the atria, and a membrane called the *interventricular septum* separates the two ventricles. Each chamber of the heart plays a specific role in pumping blood, and the anatomy of each chamber fits its function perfectly. The next section details each chamber.

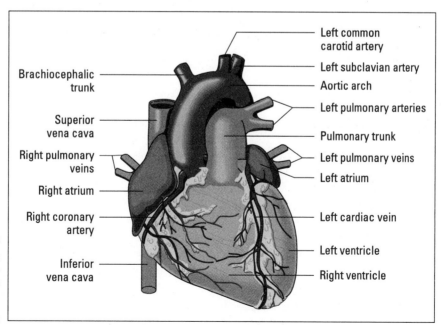

**Figure 9-1:**
Anterior
view of
the heart.

Brachiocephalic trunk

Superior vena cava

Right pulmonary veins

Right atrium

Right coronary artery

Inferior vena cava

Left common carotid artery

Left subclavian artery

Aortic arch

Left pulmonary arteries

Pulmonary trunk

Left pulmonary veins

Left atrium

Left cardiac vein

Left ventricle

Right ventricle

*From LifeART®, Super Anatomy 1, © 2002, Lippincott Williams & Wilkins*

The heart also contains several *valves,* the names of which tell you their anatomic location or characteristics (see Figure 9-2). For example, the *atrioventricular* (AV) *valves* are between the atria and the ventricles; the *bicuspid* (AV) *valve* has two flaps, and the *tricuspid* (AV) *valve* has three flaps. The *semilunar valves* are shaped like half-moons. Valves act like locks on a canal: They allow measured quantities of blood into a chamber, and they prevent blood from flowing backward. Valves keep blood flowing in the right direction on the pathway, which helps maintain the proper rhythm action in the heart and in you.

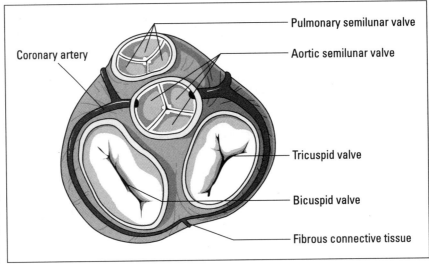

**Figure 9-2:**
Heart
valves.

Coronary artery

Pulmonary semilunar valve

Aortic semilunar valve

Tricuspid valve

Bicuspid valve

Fibrous connective tissue

*From LifeART®, Super Anatomy 1, © 2002, Lippincott Williams & Wilkins*

# Following Your Heart: Physiology of the Heart

In this section, you take a look at the functions of the heart — its ability to work as a pump and also to keep itself going by generating a heartbeat. Blood follows a set path through the body, which is described in the section, "On the beating path: The path of blood through the heart and body." The generation of a heartbeat and a pulse are described as mechanics of the heart, and you find information about them in the sections called "Generating electricity: The cardiac cycle" and "Putting your finger on your pulse."

# On the beating path: The path of blood through the heart and body

The heart is a *double pump* or a *two-circuit system* for — get this — <u>two</u> reasons. First, the heart has <u>two</u> kinds of chambers — atria and ventricles — the atria contract simultaneously, and then the ventricles contract simultaneously. Second, the heart pumps blood through <u>two</u> routes — the arterial and venous systems. See a pattern here?

The path of blood through the heart makes the anatomy of the chambers clear, and it also brings to light exactly why the heart is considered a double pump. Figure 9-3 shows you the internal anatomy of the heart as well as the path of blood through the heart.

Look at Figure 9-3 often while you read this section. Better yet, trace the path with red-and-blue-colored pencils: red for the oxygenated blood (arterial system) and blue for the deoxygenated blood (venous system).

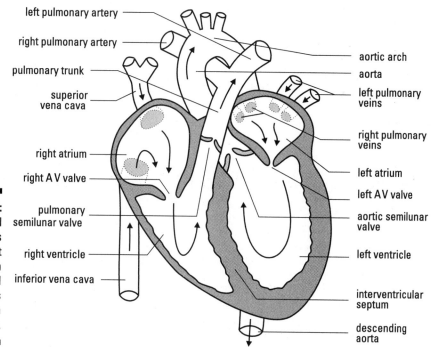

**Figure 9-3:**
Internal
structures
of the heart
and the path
that blood
takes
through
them.

left pulmonary artery

right pulmonary artery

pulmonary trunk

superior
vena cava

right atrium

right A V valve

pulmonary
semilunar valve

right ventricle

inferior vena cava

aortic arch

aorta

left pulmonary
veins

right pulmonary
veins

left atrium

left A V valve

aortic semilunar
valve

left ventricle

interventricular
septum

descending
aorta

To get through your entire body, the blood that contains oxygen must be pumped pretty forcefully out of the heart to make it through the entire path. And it is.

On its first circuit, the heart pumps oxygenated blood (the red blood with oxyhemoglobin) out through your arteries and to all the far-reaching locations around your body. After oxygen is delivered to cells and tissues, the deoxygenated blood (the purplish blood with deoxyhemoglobin) is returned to the heart by the veins. Then the blood goes through its second circuit when the deoxygenated blood is pumped from the heart to the lungs, so that it can become oxygenated again. Remember, the heart is a double pump, so there are two circuits: (1) heart to lungs, and (2) heart to body. The following numbered list shows the steps along the pathway through the heart:

1. Oxygenated blood leaves the lungs and travels into the *left atrium* via the *pulmonary veins*.

2. The left atrium contracts, thus forcing oxygenated blood through the bicuspid (left AV) valve and down into the left ventricle.

3. The left ventricle pumps oxygenated blood through the *aortic semilunar valve* into the main artery, the *aorta*.

4. Oxygenated blood travels around the body through arteries that branch off the aorta.

5. In the body's tissues, oxygen is delivered (to be used in metabolic reactions: oxygen is consumed; carbon dioxide and water are produced; see Chapter 2), and the waste product carbon dioxide is picked up in the tiniest of the blood vessels — the *capillaries*. (See the "Exchanging oxygen: Your capillaries" section coming up later in this chapter.) The blood is now *deoxygenated* because the oxygen has been delivered.

6. Deoxygenated blood returns to the heart through *veins*.

7. Before entering the heart, the deoxygenated blood passes through the largest veins in the body — the *superior vena cava* and the *inferior vena cava* — entering the *right atrium*.

8. The right atrium contracts, which forces deoxygenated blood through the *tricuspid (right AV) valve* down into the *right ventricle*.

9. The right ventricle contracts, forcing deoxygenated blood through the *pulmonary semilunar valve* and into the *pulmonary arteries*.

10. The pulmonary arteries take deoxygenated blood from the heart into the *lungs* so that it can become oxygenated.

11. The entire process — pumping of blood from the heart to the lungs, the return of blood to from the lungs to the heart, and then the pumping of blood again from the heart to the rest of the body — repeats throughout every minute of your entire life.

*Pulmonary circulation* refers to the path of blood from the heart to the lungs and back to the heart. *Systemic circulation* refers to the path of blood from the heart to the rest of the body and back to the heart. The two systems — or circuits — work together to achieve the goal of keeping your blood oxygen level within normal limits. There can't be too little oxygen in your blood at any time; nor can there be too much carbon dioxide. Homeostasis (see Chapter 8) is maintained when pulmonary and systemic circulation are intertwined.

# Generating electricity: The cardiac cycle

Whether you're aware of it, you generate electricity. Your brain generates electrical impulses, as does your heart. Within the cardiac muscle tissue are specialized fibers that make up the *cardiac conduction system,* which is responsible for generating your heartbeat. The largest mass of these specialized fibers is found at the *sinoatrial (SA) node,* also known as the *pacemaker* because it's the place where the heartbeat originates. The SA node lies on the back wall of the right atrium, near where the superior vena cava enters the heart. If your SA node stops working correctly, you need an artificial pacemaker installed on your heart. Initiating the electrical impulse that starts the beginning of the cycle, the SA node controls the *cardiac cycle.* After the SA node sends the impulse, the following events happen in the cardiac cycle:

1. **The right atrium and the left atrium contract at the same time, forcing blood into the relaxed ventricles.**

2. **After the SA node prompts the right atrium and the left atrium to contract, the electrical impulse passes to the atrioventricular (AV) node.**

   The AV node is also located in the right atrium but near the septum that divides the right atrium and left atrium from the ventricles.

3. **The AV node passes the impulse to fibers located in the septum called the AV bundle or bundle of His (pronounced "hiss").**

   The AV bundle branches twice: One branch heads right, and one branch heads left. At the ends of these branches are the Purkinje fibers, which are the fibers that deliver the impulse causing the ventricles to contract.

4. **The right atrium and the left atrium relax, and both ventricles contract simultaneously, forcing blood into the aorta.**

   Then the blood travels through systemic circulation and into the pulmonary arteries to send blood through pulmonary circulation.

5. **The right atrium, the left atrium, and the ventricles relax for less than half a second.**

The entire cardiac cycle takes about 0.85 of a second, based on the average of 70 heartbeats per minute. If your cardiac cycles take less time, your heart is beating too fast *(tachycardia);* if there is too much time between your cardiac cycles, your heart is beating too slow *(bradycardia).*

## *Putting your finger on your pulse*

Each cardiac cycle generates one heartbeat, and the pace of your cardiac cycles is your *pulse*. Normally, the heart beats about 70 times per minute (when resting), so the normal pulse is 70. Your pulse can be felt at certain spots around your body. Most commonly, the pulse is felt with any fingers (except the thumb) on the inside of the wrist or on the carotid artery of the neck. What you feel as you touch these spots is your artery expanding as the blood rushes through it. Immediately after the blood expands that spot in the artery, the artery returns to its normal size. The blood rushes through the same spot with the same regularity as your cardiac cycle, so each time you feel the artery expand, you can gauge how often your cardiac cycle is occurring and measure your pulse.

# *Looking at Your Blood Vessels*

Arteries usually have veins of the same size running right alongside or near them. Arteries and veins serve all parts of the body; for this reason, arteries and veins often have the same name. The major blood vessels of the body are diagrammed in Figure 9-4, which shows how the arterial and venous systems work together.

Think of your blood vessels as projections from your heart that wind through your entire body, much like your nerves are projections from your brain and spinal cord. This analogy really isn't far from fact. The heart and the blood vessels are lined with a continuous sheet of endothelial tissue. This means that the same tissue that lines the inside of the heart also lines the major vessels that attach to the heart: the aorta, the vena cava, and the pulmonary arteries and veins, as well as all of their branches throughout the body.

The five different types of blood vessels are *arteries, arterioles, capillaries, venules,* and *veins.* The arterial system (arteries and arterioles) decrease in size as they spread throughout the body. Arteries carry oxygenated blood away from the heart and to the body; arterioles are smaller branches off of arteries. Arterioles also carry oxygenated blood, usually from arteries to capillaries. The venous system (veins and venules) is opposite of the arterial system. The smaller venules carry deoxygenated blood first, from a capillary to a vein. Then, the larger veins carry the deoxygenated blood from the venules of the body back to the heart and lungs. Capillaries are the tiniest blood vessels that serve as a bridge between the vessels carrying oxygenated blood and those carrying deoxygenated blood — usually the arterioles and venules. (See the "Exchanging oxygen: Your capillaries" section in this chapter.)

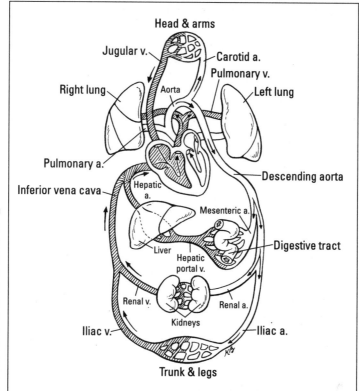

**Figure 9-4:**
Pulmonary
and
systemic
circulation
working
together
through
arterial and
venous
systems.

Head & arms

Jugular v.

Carotid a.

Pulmonary v.

Right lung

Aorta

Left lung

Pulmonary a.

Descending aorta

Inferior vena cava

Hepatic a.

Mesenteric a.

Digestive tract

Liver

Hepatic portal v.

Renal v.

Renal a.

Kidneys

Iliac v.

Iliac a.

Trunk & legs

## Analyzing arteries and veins

Anatomically, arteries have thick walls. The innermost layer of an artery —
called the *tunica interna* — is made of endothelium, just like the inside of
the heart. Covering the endothelium is a layer of elastic tissue, which allows
the artery to expand when necessary. Covering the layer of elastic tissue is
a thick layer of smooth muscle, which makes the artery constrict when the
muscle contracts. The elastic tissue and smooth muscle layers form the
middle part of the artery — the *tunica media.* The outermost layer of an
artery — the *tunica externa* — is a thick layer of connective tissue.

Veins have a similar anatomy, although their walls tend to be thinner than
arteries. The tunica interna of a vein is also made of endothelium, and the
tunica media has a layer of elastic tissue covering the endothelium as well
as a layer of smooth muscle. However, the smooth muscle layer of a vein is
much thinner than that of an artery. The reason for this is that blood passes

through veins in a manner different than it does in arteries. Because arteries receive blood being pumped from the heart, there usually is sufficient pressure in the arteries to move blood through the arteries and arterioles. Sometimes, though, the arteries need a little squeeze to increase the pressure and force the blood through, which is why there is a larger muscular layer in an artery. The pressure in capillaries is low, and then the blood moves into the venous system, where there is virtually no blood pressure.

Blood moves through veins as a result of contraction of skeletal muscles. As you move your arms, legs, and torso, your muscles contract, and those movements "massage" the blood through your veins. The contraction of skeletal muscle provides a type of kneading action to squeeze the blood through. However, the blood moves through the vein a little bit at a time. There are valves inside veins that keep blood from flowing backward. The valves open in the direction that the blood is moving, and then shut once the blood passes through to keep the blood heading toward the heart. All veins eventually lead to the main veins — the *superior vena cava* and the *inferior vena cava.*

The *inferior vena cava* (the lower division of the vena cava that enters the bottom of the heart) is formed when the *common iliac veins* merge. The common iliac veins are formed by the *internal iliac vein,* which returns blood from the pelvic organs, and the *external iliac vein,* which returns blood from the legs. Likewise, the *renal veins* return blood from the kidneys, and the *hepatic veins* return blood from the liver up to the heart through the inferior vena cava. From the head and arms, deoxygenated blood returns to the heart through the *jugular vein,* which connects to the *superior vena cava* (the top portion of the vena cava that enters the top of the heart) by way of the *brachiocephalic vein.* The veins of the arm — the *ulnar vein, radial vein,* and *subclavian vein* — all lead to the brachiocephalic vein.

When oxygenated blood leaves the heart via the aorta, blood immediately travels up through the *carotid artery* to serve the head or down through the *descending aorta* to serve the abdominal organs and lower limbs. The descending aorta branches into the *mesenteric artery,* the main artery into the digestive tract, the two *renal arteries,* which supply the kidneys, and the *common iliac artery,* which supplies the pelvis and legs.

## Exchanging oxygen: Your capillaries

*Capillary exchange* (see Figure 9-5) is the term used to describe how oxygen and nutrients get from the bloodstream to the cells and how carbon dioxide and waste materials get from the cells to the bloodstream.

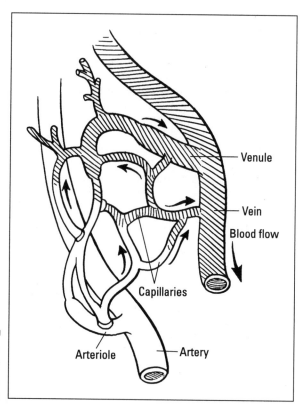

**Figure 9-5:**
Capillary
exchange.

Oxygenated blood passes through the arteries and branches into the arterioles. After passing through the arterioles, blood enters the smallest branches: the capillaries. The conglomeration of capillaries that lie between larger blood vessels is a *capillary bed* (refer to Figure 9-5). A capillary bed connects the arterial system with the venous system; specifically, capillaries form a bridge between the arterioles and the venules. Capillary beds are everywhere in your body. This is why you bleed anywhere that you even slightly cut your skin. But because of their omnipresence, capillaries are able to provide the important materials of life to all your cells in all your tissues.

The cell wall of a capillary is only one-cell thick. So because so many capillaries spread throughout your entire body, the thin capillaries come in contact with all your tissues. When blood passes through the capillary, the oxygen that's present in blood from the nearby arteriole diffuses across the thin wall of the capillary directly into the tissue fluid. (See Chapter 3 for more on diffusion.) Tissue fluid surrounds all the cells in a tissue, which explains why your skin looks wet and moist when you injure it. The tissue fluid acts as a medium through which material can pass into the cells. Oxygen naturally occurs as a gas, and the erratic movement of gas wouldn't ensure that all cells would get the needed oxygen. So to better control the movement of oxygen molecules, oxygen needs to be dissolved in a watery solution. The liquid component of blood and tissue fluid serve that purpose. With oxygen dissolved in the tissue fluid, it can move from the capillaries to the tissue fluid and then diffuse into the cells of your body.

At the end of the capillary near the arteriole, the oxygen (and nutrients such as water, glucose, and amino acids) diffuse into the tissue fluid. At the end of the capillary closer to the venule, the carbon dioxide and other waste materials diffuse out of tissue fluid and across the capillary's membrane into the blood. Then, the blood continues on through the venous system, and those waste materials are deposited in the proper locations on their way out of the body. For instance, the carbon dioxide diffuses out of the bloodstream in the lungs, so it can be exhaled and thus removed from the body.

A capillary, as tiny as it is, really is the goal for blood to reach. Blood cannot serve its purpose of exchanging oxygen and nutrients for carbon dioxide and waste anywhere but in the capillaries, so the blood has to get there. Capillaries also do more than just exchange the good and bad substances. They also play a role in blood pressure. (See the "Exercise lowers blood pressure" sidebar in this chapter.)

# Pathophysiology of the Circulatory System

Sometimes in the stream of life, boats sink and logs jam. When things are flowing smoothly, life is wonderful; but when the flow is blocked, life's a you-know-what. The same holds true for the circulatory system. When your circulatory system is working at its peak and your cells are flowing freely through your body, your health is great. But, when a problem develops in the blood cells or when the blood flow is reduced because of a blockage, your health suffers. In this section, you'll find out about what causes some of the problems in the circulatory system. High blood pressure, heart disease, and stroke are the biggies, the ones that most commonly cause health problems and death in Western societies, usually as a result of lifestyle (diet, lack of exercise, stress). But, there are also genetic problems that affect the circulatory system. Sickle cell anemia is one such genetic problem, and information on it also is included in this section.

# High blood pressure

*Blood pressure* is a term used to describe the force that blood has when it pushes against the wall of an artery. Two factors affect the force that blood has: *cardiac output* and *peripheral resistance.* Cardiac output, just like the name sounds, is the amount of blood the heart pumps out. Peripheral resistance describes the amount of (or lack of) "give" in the vessel walls. The higher the cardiac output — that is the higher the heart rate or the more blood there is for the heart to pump — the higher the blood pressure must be to get the job done. Likewise, the more resistant the arteries are to expanding as the blood rushes through, the higher the blood pressure becomes.

Several factors are involved in the increase of blood pressure. First, the *fight-or-flight response* results in a release of the hormone *adrenaline* (epinephrine), which increases the heart rate (see Chapter 8). Other emotional stimulation, such as sexual contact or anxiety, also can cause stimulation of nerves that send an impulse to increase the heart rate. An increase in heart rate causes an increase in cardiac output, which increases blood pressure. An excess of an enzyme called renin (the excess is usually caused by a *kidney problem*) can cause imbalance in a group of enzymes and hormones called the renin-angiotensin-aldosterone system, which causes absorption of sodium.

Too much sodium causes arterioles to constrict, which raises the blood pressure. *Antidiuretic hormone* (ADH; see Chapters 8 or 12) causes water to be reabsorbed from the kidney back into the blood stream, which increases the blood volume. The increase in blood volume raises the blood pressure. Some of these actions are normal and healthy ways of self-preservation or homeostasis. However, when the excess of sodium comes from the diet or there is a hormonal problem resulting in too little or too much of key hormones being secreted, the blood pressure can become quite high, sometimes requiring medical intervention to bring it back into the normal range.

However, the body does try to normalize the blood pressure, too. As part of homeostasis, receptors in the arteries detect the pressure. An impulse is then sent from these *pressoreceptors* to the medulla oblongata of the brain (see Chapter 7), informing the brain of the blood pressure. If the blood pressure is above normal, the brain sends out impulses to cause responses that decrease the heart rate and dilate the arterioles, both of which decrease blood pressure.

*Hypertension* (high blood pressure) is any systolic measurement higher than 140 millimeters of mercury (abbreviated mm Hg; Hg is the abbreviation for the element mercury) or a diastolic measurement higher than 90 mm Hg. The blood pressure, which is measured in an artery, tells you what the pressure is

## Exercise lowers blood pressure

You can control your blood pressure, but it does require some physical work. Whether you call it work or exercise (I prefer to look at it as "playtime"), moving your body is extremely beneficial. Besides strengthening your skeletal muscles, exercise strengthens your cardiac muscle and bones. It also can lower your blood pressure.

You know how you have many, many capillary beds all over your body? Well, not all of them are open at the same time. If you're not moving, your capillary beds close. *Sphincters* inside the capillaries tighten to prevent blood from flowing into the capillary bed. (Blood then is shunted from an arteriole directly to a venule through a nearby *arteriovenous shunt.*) When no blood is in the capillary bed, diffusion of oxygen or nutrients to the tissue fluid can't happen. And without that action happening, metabolic activity is lessened. However, when you're exercising or playing, the sphincters inside the capillaries relax, opening the capillary bed to blood flow and metabolic activity. The more capillary beds that are open, the more room that blood has to flow through the body. With more vessels open, the lower the peripheral resistance and the lower the blood pressure. Now, go play!

when the left ventricle contracts (*systolic,* top number) and when it relaxes (*diastolic,* bottom number). A normal blood pressure is around 120/80 mm Hg. The higher the systolic and diastolic values, the more pressure there is on the walls of the arteries, and the harder the heart is working.

Many factors can cause high blood pressure. Arteries are lined with endothelium, and the tissue can change over time. Sometimes other diseases (such as diabetes, kidney disease, or thyroid disease) can cause vascular changes, such as in the lining of the blood vessels. Lifestyle factors — such as too much dietary sodium or fat, little or no exercise, obesity, stress, smoking, or oral contraceptive use — play a big role in the development of hypertension.

If the blood pressure is constantly too high, the continually increased pressure eventually damages the inside of the blood vessels. When tissues, such as the blood vessels, are damaged, platelets rush to the site of injury. The platelets begin to congregate at the injured sites along the blood vessels, as do proteins involved in clotting (such as fibrinogen). If a clot (*thrombus*) forms in an artery, it can block the artery. Also, a clot can dislodge and travel to another site in the body (then called an *embolism*). If a clot dislodges from a blocked artery in, say, your leg, and then travels to and lodges in a lung, the clot is a *pulmonary embolism,* which can be a fatal problem.

Even if clots don't form, blockages can be caused by the accumulation of platelets and fibrin, and swelling (*edema*) around the blocked artery can occur from plasma seeping through the damaged vessel. As the inside *(lumen)* of the blood vessel narrows, the peripheral resistance increases, and that increases blood pressure. Blood pressure can be reduced with lifestyle changes (see the "Exercise lowers blood pressure" sidebar in this chapter) and medications such as *beta-blockers,* which dilate the arteries.

# Heart disease and stroke

Another common disease that results from narrowing arteries is *atherosclerosis,* also called "hardening of the arteries." Some narrowing and loss of expandability in the arteries occurs with aging, but when it happens too fast or too soon, the risk of *coronary artery disease* (CAD) and *myocardial infarction* (MI) increases greatly. In people with atherosclerosis, fats begin to build up on the endothelium lining the vessels. Then, fibrous plaques begin to form on the fatty deposits. Next, calcium deposits become enmeshed in the fibrous plaques. At this point, the blood vessel is pretty well blocked, and those blockages can cause reduced blood flow to tissues and organs (called *ischemia*), as well as sudden pain as blood cells try to squeeze through (called *angina*). With reduced blood flow comes oxygen and nutrient deprivation, which can cause metabolic problems and symptoms such as fatigue.

If blood flow is reduced to the arteries that supply the heart with oxygen and nutrients, that is, the coronary arteries, a *myocardial infarction* (heart attack) can occur. During a heart attack, the lack of oxygen causes some of the tissues of the heart to die. In a massive heart attack, a large percentage of the heart's tissues die, and this usually leads to death.

If blood flow is reduced to the arteries that supply the brain with oxygen and nutrients (the internal carotid arteries and the cerebral arteries), a *cerebrovascular accident* (CVA), more commonly called a *stroke,* occurs. When the oxygen supply to the brain is cut off during a stroke, some of the brain tissues die. Symptoms of a stroke depend on where in the brain the affected tissues are. If the tissue that dies is in the center for speech, then speech is slurred. If the tissue that dies is in the center for vision, then the stroke victim might become blind. Besides blocked arteries, an *aneurysm,* an artery that bursts, can also cause strokes. Just as increasing water pressure behind a dam can cause the dam to break, the increased blood pressure behind the blockage can cause an artery to rupture.

Although some risk factors for heart disease and stroke cannot be prevented, such as increased age, gender, or race, other risk factors can be prevented. The best ways to decrease your risk of being yet another heart disease statistic is to do the following:

✔ Make sure you never smoke (or quit smoking).

✔ Limit the amount of fat and sodium in your diet by making wise food choices (plenty of fruits and vegetables, whole grains, low-fat dairy products, and lean protein sources, such as chicken, fish, nuts, soy); go light on heavily processed foods, which tend to have too much sugar, salt, or fat and little nutrient value.

✔ Make exercise a part of your daily routine to ensure that your blood keeps flowing through the blood vessels. This can keep plaques from forming and reduce blood pressure by opening up those capillary beds. (Refer to the "Exercise lowers blood pressure" sidebar in this chapter.) This seems akin to removing the hydrocarbons from a car's exhaust system.

✔ Drink the amount of water that you need to stay well hydrated (usually 64 ounces per day) because dehydration can increase blood pressure. Think of this as a "fill and flush" (rather than a flush and fill) for *your* engine. (Fill and *flush* is pretty appropriate after drinking all that water, don't you think?)

✔ Try to reduce the stress in your life so that your nervous system is not flooding your body with hormones that increase the heart rate and blood pressure for no good reason. That is, don't get upset over the little things; in other words, save stress reactions for occasions when your life is truly threatened or you are genuinely scared.

## Sickle cell anemia

Sickle cell anemia, an inherited disorder, results in the sudden change of normally shaped blood cells to oddly shaped blood cells. Normally, red blood cells are shaped like flying saucers, but in this disorder, the red blood cells curve, like a sickle. (Remember the old hammer and sickle flag of the former USSR? I think that they look like boomerangs myself. But "boomerang cell anemia" just doesn't have the same ring.)

Sickle cell anemia occurs most often in Africans and people of African descent. The gene for this disease causes affected people to produce defective hemoglobin molecules called *hemoglobin S.* People in Africa (and because the disease is genetic, those of African descent) have such a high incidence of this gene because hemoglobin S makes people resistant to the mosquito-borne disease *malaria,* which is so prevalent in Africa. So, what is viewed as the production of a defective molecule actually was a genetic change that protected a population.

Sickled red blood cells have a shorter life span than regular RBCs, and they're unable to carry as much oxygen through the blood stream. These cells also undergo *hemolysis* (destruction of RBC) at a higher rate than do regular RBCs. Their odd shape causes the sickled cells to get stuck in small blood vessels, such as capillaries. When this happens, the sudden blockage causes pain and swelling, and the tissues in the area can die because of a lack of oxygen or build-up of waste products that cannot be removed from the body; all these events are part of a *sickle cell crisis.* Several factors, such as cold temperatures, stress, infection, even high altitudes or low levels of oxygen in the bloodstream, can trigger a sickle cell crisis.

The end result of a sickle cell crisis is that tissues of the body die, and organs, such as the kidneys, can be permanently damaged. Over time, as a patient experiences more sickle cell crises, the damage can be overwhelming. People with sickle cell disease have shortened life spans, although not as short as in the past. Now, instead of a life expectancy of fewer than 30 years, blood transfusions and medications have extended the life spans of those with sickle cell anemia into their 40s or 50s.

# Chapter 10

# Breathe a Sigh of Relief: The Respiratory System

. . . . . . . . . . . . . . . . . . . . . . . . . . . . . . . . . . . . . . . . . . . . .

## In This Chapter

▶ Checking out the parts of the respiratory system

▶ Finding out the difference between breathing and respiration

▶ Understanding what the respiratory system does

▶ Seeing the havoc that smoking can wreak on your body

. . . . . . . . . . . . . . . . . . . . . . . . . . . . . . . . . . . . . . . . . . . . .

*W*ell, you're alive and reading this page, so I can safely assume that you're a living, breathing human. But do you know how your breathing keeps you a living human? Without your respiratory system working properly, you'd be unable to breathe — you'd be dead. You can't live without your respiratory system, so respect it, find out what's in it, what it does, and how you can keep it healthy.

This chapter clarifies the parts and activities of your respiratory system — that organ group that provides you with oxygen. Your respiratory system oversees a number of vital body functions, including:

✔ The movement of air in and out of your lungs (ventilation).

✔ The exchange of oxygen and carbon dioxide between the air in the lungs and the blood.

✔ The exchange of oxygen and carbon dioxide between the blood and the cells of the body.

✔ The metabolism of oxygen, which results in the production of carbon dioxide, in your cells (see Chapter 2).

# Knowing Your Respiratory Anatomy

Considering the enormously important job that this system has, the "parts" of the respiratory system are few in number and pretty simple. Your nose, pharynx, trachea, lungs, and diaphragm are the major parts of the respiratory system. You also have a few accessory organs — those that work with the major parts to perform a related function. An example of this is your vocal cords, which work with your pharynx and air taken in through the nose to produce your voice. For a visual representation of the key organs that keep you breathing, see Figure 10-1.

## Nosing around

This is one time that it's okay to turn your nose up at your anatomy. Really. Point your nose up while looking in the mirror. (Yes, you have to put the book down.) See the two big openings? Those are your *nostrils,* and that's one of two places where air enters and exits your respiratory system. Now, see all those tiny hairs in your nostrils? Those little hairs serve a purpose. They not only keep older men busy with some trimming, they also trap dirt and dust particles to keep tiny nasty bacteria that can make you sick from going into your respiratory system. Okay. You can put your head down, now. The rest of your respiratory parts are way inside of your body, so you can't see them in your bathroom mirror.

Just beyond your nostrils, the *nasal septum* separates your *nasal cavities.* Inside the nasal cavities, the *nasal conchae,* three tiny bones, provide more surface area inside the nose because they're rolled up (like conch shells). The skin on the nasal conchae and lining the inside of the nasal cavity (called the *mucous membrane*) produces mucus to provide warm moisture to the air entering the body (kind of like your own personal humidifier). Covering the membrane, tiny *cilia* help to filter the incoming air.

In the corners of the eyes, the *lacrimal glands* secrete tears that drain through *tear ducts* into the nasal cavities. This is why your nose runs when you cry.

Your *sinuses* are air spaces in your skull that serve to lighten the weight of your head. The sinuses open into the nasal cavities so they can receive air as you breathe, and, like the nasal cavities, the sinuses are lined with mucous membranes. Any mucous membrane can become infected or inflamed. So, when infection or inflammation extends beyond your nasal cavities and winds up in your sinuses, the condition is *sinusitis.* Sinus headaches occur when the passage to a sinus is blocked with mucus or inflamed tissue. Ouch!

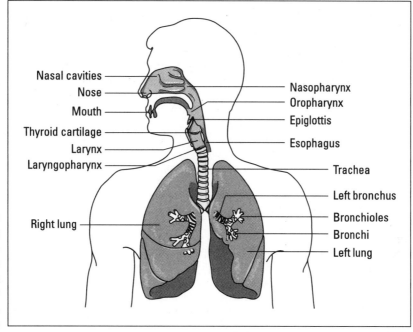

**Figure 10-1:**
Anatomic
structures
of the
respiratory
system.

Nasal cavities
Nose
Mouth
Thyroid cartilage
Larynx
Laryngopharynx
Right lung

Nasopharynx
Oropharynx
Epiglottis
Esophagus
Trachea
Left bronchus
Bronchioles
Bronchi
Left lung

*From LifeART®, Super Anatomy 1, © 2002, Lippincott Williams & Wilkins*

## Swallowing some facts

This fact shouldn't be hard to swallow: Just as food passes through your *pharynx* (throat) on its way to your stomach, air passes through your pharynx on its way to your lungs. Along the way, it passes through and by some other important structures, like your larynx and tonsils.

Your pharynx is divided into three regions based on what structures open into it:

- ✔ **Nasopharynx:** The top part of your throat where your nasal cavities drain.

- ✔ **Oropharynx:** The middle part of your throat into which your mouth opens.

- ✔ **Laryngopharynx:** The lower part of your throat where your *larynx* (voice box) resides.

## Why food and drink can spurt out of your nose

Have you ever laughed so hard while eating or drinking that whatever was in your mouth comes exploding out through your nostrils? If you've ever wondered how that can happen, this is your sidebar. If you press your tongue to the roof of your mouth, you can feel your *hard palate*. This bony plate separates your mouth (oral cavity) from your nose (nasal cavities). If you move your tongue backwards along the roof of your mouth, you reach a soft spot. This spot is the *soft palate*.

Beyond the soft palate is where your nasal cavities drain into your throat; this area is your *nasopharynx*. Your soft palate moves backward when you swallow so that the nasopharynx is blocked. Normally, the soft palate blocking the nasopharynx keeps food from going up into your nose. But when you're laughing and eating or drinking at the same time, your soft palate gets confused. When you go to swallow, it starts to move back; but, when you laugh suddenly, it thrusts forward, allowing whatever is in your mouth to flow up into your nasal cavities and immediately fly out of your nostrils to the delight of everyone around you.

Although many people call the *larynx* a voice box, it's shaped like a triangle. At the tip of the triangle is *thyroid cartilage,* commonly known as your "Adam's apple." If you could look down your throat onto the top of your larynx, you could see your *glottis,* the opening through which air passes. When you swallow, a flap of tissue called your *epiglottis* covers your glottis and blocks food from getting into your larynx. If food does "go down the wrong pipe," so to speak, you cough until it's forced back out of your larynx.

Inside your glottis, your *vocal cords* vibrate when air passes over them. As they vibrate, they produce sound waves known as your voice; the more they vibrate, the louder your voice sounds. The vocal cords themselves are just gathered mucous membranes covering ligaments of elastic tissue. (For more on ligaments, see Chapter 4.) Tightening your vocal cords can change the pitch of your voice. When you tighten them, the glottis narrows, and your voice sounds higher.

As I write this, my little girl is singing at the top of her 2-year-old lungs. Her extremely tight young vocal cords are vibrating beyond belief; that is, she's loud and high pitched (but oh-so-cute).

# Tracking your trachea into your lungs

Your *trachea* (windpipe) is the tube that runs from your larynx to just above your lungs. Just behind your sternum (see Chapter 4), your trachea divides into two large branches called *bronchi* collectively — *bronchus* in the singular. Each bronchus enters a lung, and then splits into smaller bronchi (secondary and tertiary branches). The tertiary bronchi branch into still smaller tubes called *bronchioles,* which are akin to twigs on a small branch. At the end of the smallest bronchioles are little structures that look like raspberries. These little berry-shaped air sacs are *alveolar sacs,* and each sac contains many *alveoli* (*alveolus* is singular). Each of the approximately 300 million alveoli in your lungs is wrapped with capillaries.

Although the alveoli are the smallest parts of the respiratory system, they're the most numerous and most important because they are involved in gas exchange. Oxygen moves from the alveoli in the lungs to the capillaries surrounding the alveoli so it can be transported through the bloodstream throughout the body. Carbon dioxide moves from the capillaries surrounding an alveolus into the alveolus so it can be exhaled from the lung. Together, the alveolar wall and its abutting capillary wall make up the *respiratory membrane* where this gas exchange occurs. (See the section on "Going through the Respiratory Membrane.")

# Lining your lungs

The moist *parietal membrane* "glues" the lungs to the borders of the thoracic cavity (see Chapter 1). The *visceral membrane* encases the lungs tightly and surrounds the *mediastinum,* the region that separates the left and right lungs and houses the heart, thymus, and part of the esophagus. Together, the parietal and visceral membranes are the *pleural membranes.*

Between the parietal and visceral membranes, is a thin *interpleural* (meaning "between the pleura") *space.* It's filled with fluid that keeps the tissues of the parietal and visceral membranes moist. Because this fluid has a lower pressure than the pressure in the atmosphere (that is, little air is in fluid), your lungs stay inflated. If air gets into one of your interpleural spaces, the pressure changes and your lung collapses. Then your respiratory membranes can't provide oxygen to your blood. If both of your lungs were to collapse, you could suffocate quickly.

# Breathing Some Life into Your Body

Respiration means many things to the proper functioning of your body — all involving the supply of oxygen that you need to stay alive.

*Ventilation* — the act of breathing — isn't the same as *respiration* — the exchange of gases and nutrients at the cellular level. You don't have to think about performing either task, but ventilation involves the physical movement of your diaphragm muscle, rib cage, and lungs. Respiration is mainly a series of metabolic reactions (see more on cellular respiration in Chapter 2) that happen without you noticing any physical movement.

*Inspiration* is the action of breathing in, taking air into the lungs (see Figure 10-2). At the moment that you breathe in, all of your air passages — that is, all bronchi, bronchioles, and alveoli — are open. (Refer to the section "Tracking your trachea into your lungs" earlier in this chapter for an explanation of these terms.) Envision the air that you breathe in as mist. When you inhale, the mist fills your nasal cavities, pharynx, and all air passages, all at the same time. Your body basically creates a vacuum that sucks in air. Imagine how the leather of a fireplace bellows sticks together when all the air is squeezed out and how air fills the bellows as the handles are moved farther apart. Moving the handles apart creates a vacuum that sucks the air in. Your lungs work in a similar way, and your brain controls the entire process.

In the part of your brain called the medulla oblongata (see Chapter 7 for more on the brain) lies a group of nerve cells that emit impulses that inspire you to take a breath. These special neurons are your *respiratory center*. The following numbered list shows how you get from the condition that prompts the impulse to actually taking a breath:

1. When special chemoreceptor cells (see Chapter 7) in your carotid arteries and aorta detect a higher level of carbon dioxide than oxygen or a level of hydrogen ions that's too high (too much hydrogen signals an acidic condition), they alert your respiratory center to the problem. (*Ions* are atoms of an element with a positive or negative charge. For more on ions, see Chapter 5.)

2. After the cells in your respiratory center hear the alarm, so to speak, they send an impulse to your diaphragm muscle and the muscles surrounding your rib cage to contract.

   The diaphragm muscle at the bottom of your thoracic cavity is responsible for pushing air out of your lungs. (Refer to Figure 10-2.)

3. When the muscles around your rib cage (the external intercostal muscles) contract, your rib cage moves upward and outward.

4. At the same time that your rib cage moves up and out, your diaphragm muscle contracts (moves downward), enlarging your thoracic cavity much like you widen the fireplace bellows when you move the handles apart from one another.

5. As your thoracic cavity enlarges, the air pressure inside the thoracic cavity decreases. This creates a vacuum that causes air to rush in through your nose (and mouth, if it's open). As air rushes in, the lungs expand and the alveoli within the lungs enlarge.

6. After the air comes in, your respiratory center stops sending the impulse to contract your rib cage and diaphragm.

7. When the impulses stop, your diaphragm relaxes and moves back up (because of abdominal muscle tone), and your rib cage relaxes and moves inward and downward (mostly because of gravity).

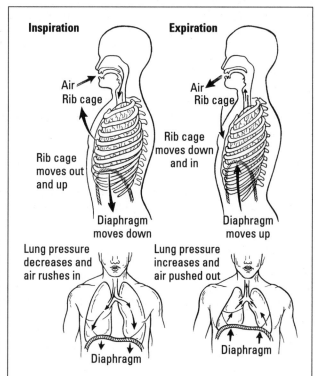

**Figure 10-2:**
The inspiration (inhalation) and expiration (exhalation) process.

COOL BODY BITS

## Taking measured breaths

To determine whether you're breathing properly and if your lungs are functioning well, physicians can measure the amount of air you inhale and exhale, as well as measuring how much air remains in your lungs after you exhale and how much air you're capable of holding.

People normally inhale and exhale a total of 500 milliliters of air with each breath. Of course, if you breathe deeply, you can inhale and exhale more. This greater amount of air that you take in and expel is your *tidal volume* (TV). The maximum volume of air that you can move through your lungs in one breath is your *vital capacity* (VC). By breathing really deeply, people usually can reach a VC of 3,100 milliliters, far beyond the usual 500 milliliters that goes in and out of the lungs in one effortless breath. The VC minus the TV gives you the IRV, *inspiratory reserve volume.* This value tells you how much more

room you could make in your lungs above and beyond a normal inhalation.

You can also measure your exhalations. Your *expiratory reserve volume* (ERV) measures how much more air that you can force out beyond what you exhale during a normal breath (that is, above and beyond the exhalation part of your tidal volume). The average ERV is about 1400 milliliters. So, the VC = TV + IRV + ERV. Okay?

Then, even after you've forced out all the air you can, you still have a little more left in your lungs — the *residual volume* (RV), 1,000 milliliters on average. And some of the air you breathe in never actually gets down into your lungs. This volume, called *dead space,* is the part of the air column that fills your pharynx, trachea, and nasal cavities — places where gas exchange doesn't take place.

These actions increase the air pressure inside your lungs, which forces air out of the lungs and through your nose or mouth. Lo and behold, you're exhaling. When you're squeezing the two sides of a fireplace bellows together, you're increasing the air pressure inside the bag, which forces air out of the tip. The mechanism of exhalation operates in the same way.

# *Going through the Respiratory Membrane*

Your respiratory system would be useless without the respiratory membrane. What would be the point of inhaling and exhaling if your body couldn't do anything with the gases it took in? And, yes, air is a gas. The air that you breathe in contains mostly oxygen, but other elements are in there, too. Nitrogen, hydrogen, and sulfur all are found in air.

The respiratory membrane is not really one true membrane; it's just a term given to the two membranes where gas exchange takes place. (Refer to the

"Tracking your trachea into your lungs" section in this chapter for more on the respiratory membrane.) That is, the cell membrane of an alveoli and the membrane of the capillary associated with it exchange oxygen (that is, blood becomes oxygenated) and carbon dioxide (see Figure 10-3).

At the respiratory membrane, gases are exchanged in as quick as one or two seconds. In that brief time, this is what happens:

1. Carbon dioxide and water diffuse across the capillary's membrane into an alveoli in the lung. (See Chapter 3 for more on diffusion.)

2. Oxygen diffuses from the alveoli of the lung into the capillary, where the iron in the hemoglobin of red blood cells binds with the oxygen immediately. (See Chapter 9 for more on hemoglobin.)

3. Oxyhemoglobin forms in the red blood cells, giving the now-oxygenated blood its bright red color. (See Chapter 9 for more on oxyhemoglobin.)

4. The blood temperature in the capillary decreases slightly after oxyhemoglobin forms.

   This seemingly trivial event keeps the body temperature from increasing out of the normal range — an important part of homeostasis (see Chapter 8).

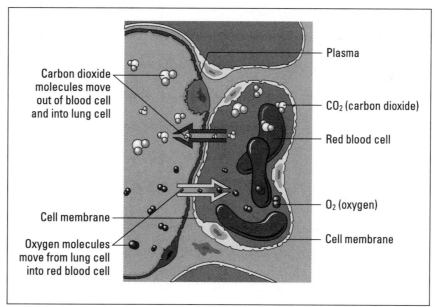

**Figure 10-3:** Oxygenation of blood at the respiratory membrane.

Plasma

Carbon dioxide molecules move out of blood cell and into lung cell

$CO_2$ (carbon dioxide)

Red blood cell

Cell membrane

Oxygen molecules move from lung cell into red blood cell

$O_2$ (oxygen)

Cell membrane

Now the blood in the capillary is oxygenated. To do the cells any good, however, oxygen needs to move from your lung back to your heart so it can be pumped into your arteries (see Chapter 9) and spread throughout your body.

# Exchanging Gases between Blood and Cells

After the oxygenated blood is pumped through your arteries, it travels to your capillaries again. But this time, the blood is in any capillary anywhere in the body. While the blood is creeping through your capillaries, the oxygen it contains diffuses out of the capillary and into the surrounding tissue fluid. At the same time, carbon dioxide diffuses out of your cells and tissues and into the surrounding tissue fluid. From there, it diffuses into the capillaries and is whisked on its way out of your body. (For more on diffusion, see Chapter 3.)

Your cells use oxygen for aerobic respiration (see Chapter 2). Oxygen and carbon dioxide are diffused from capillary to cell and from cell to capillary because oxygen and carbon dioxide reach certain high or low levels. The cell uses oxygen, and when it runs low on oxygen, it takes it from the red blood cells in the nearest capillary that have a higher concentration of oxygen. Likewise, the capillary has a low level of carbon dioxide, but the body cell has a higher level of carbon dioxide (because carbon dioxide is a byproduct of the aerobic respiration that's going on in your cells). So the carbon dioxide diffuses from the place of higher concentration to that with a lower concentration (cell to capillary).

In the blood, carbon dioxide is transported in several different ways.

- ✔ Some hemoglobin takes up carbon dioxide to form *carboxyhemoglobin*. The carboxyhemoglobin is carried through your blood like oxyhemoglobin is, and then carbon dioxide is released when the blood reaches an alveolus.

- ✔ A small percentage of carbon dioxide simply dissolves in your plasma and is carried through your bloodstream to an alveolus.

- ✔ Some carbon dioxide combines with water to form bicarbonate. Bicarbonate is an acid salt but not as toxic to cells as carbon dioxide is, so it's safer for your blood to "carry" carbon dioxide this way. When the blood reaches the lungs, the bicarbonate splits apart, releasing carbon dioxide and water. (This is the reason that your breath is moist.)

When a body cell has used up most of its oxygen and has produced carbon dioxide as a waste product, the blood comes through the capillary, and the oxygen and carbon dioxide quickly change places. The cells get the oxygen they need, and the blood becomes deoxygenated.

The deoxygenated blood (blood without oxygen) continues flowing from the capillary into your venous system. When the deoxygenated blood arrives back at your heart, it's pumped into your lungs, flowing all the way into the capillaries located there.

At this point, the process starts over again with the oxygen diffusing from the alveoli into the capillaries and the carbon dioxide diffusing from the capillaries into the alveoli. The blood is oxygenated again and sent back to the heart to be pumped through your body (systemic circulation). The carbon dioxide in the alveoli gets pushed out of your lungs on the next exhalation and is released to the outside air. Voila!

# Exploring What Can Go Wrong

You breathe in, you breathe out. It just happens, so it doesn't seem that hard. Right? Well, your body has an amazing sense of timing and rhythm, however, sometimes the body does skip a few beats or dance to a different drummer. Sometimes illness sets in and can throw your body's rhythm off, adversely affecting your respiratory system. Respiratory diseases and disorders are wide and varied. I've outlined a few of them for you in this section.

## Asthma

Asthma is probably one of the most common of all diseases, let alone respiratory diseases. More and more people are being diagnosed with asthma, and air pollution seems to be linked to it. Most of those people being diagnosed with asthma are children.

Asthma is classified as a chronic disorder involving a reactive airway. That means that the airway or airways, such as bronchioles or bronchi, occasionally become reduced in size or blocked when they react to certain triggers. The bronchial tubes can spasm, excess mucus can be secreted or the mucosal membrane lining the tubes can swell. Any of these reactions can block your airway.

Signs and symptoms of *mild* asthma include coughing, shortness of breath, or wheezing briefly after exercise or suffering those symptoms once or twice a week without exercise inducing them. Signs and symptoms of *moderate* asthma include shortness of breath, trouble breathing while resting, and an increased respiration rate (like the person is trying harder to get air into the lungs). Moderate asthma attacks can last several days. *Severe* asthma usually results in obvious wheezing and respiratory distress, and low oxygen levels in your blood *(hypoxemia)*.

You can take medications to reduce the inflammation in your airways and to prevent or reduce the allergic-type reactions that cause the swelling. If necessary, severe asthma patients may need to receive oxygen to increase the blood's oxygen level. If respiratory failure occurs, a mechanical respirator may be necessary to help the patient recover from the attack.

Asthma's two types — intrinsic and extrinsic — refer to the factors that bring on an attack.

### Extrinsic asthma

Factors outside the body bring on extrinsic asthma. Triggers often include

- Cats
- Pollen
- Dust
- Mold spores
- Outdoor air pollution
- Indoor pollution (candle soot or cigarette smoke)

### Intrinsic asthma

Internal factors — conditions that exist inside your body, not allergens — bring on intrinsic asthma. This type of asthma usually begins after a severe infection in your respiratory tract, but other triggers include

- Hormonal changes
- Environmental changes (such as heat and humidity)
- Strong fumes such as from cleaning products
- Stress
- Fatigue

As with most diseases, it's better to try to prevent the disease than to treat it. So, when possible, avoid the triggers to prevent an extrinsic asthma attack.

# Bronchitis

Bronchitis is inflammation of the bronchi. *Acute bronchitis* can occur after a respiratory infection or after exposure to irritating substances or to cold temperatures. If a virus or bacteria has spread from your nasal cavities down to your bronchi, your body produces mucus, causing a persistent cough. This type of bronchitis either clears on its own or is eliminated with antibiotic treatment.

*Chronic bronchitis,* however, is not necessarily caused by an infection. Long-term exposure to irritating substances (such as chemicals or cigarette smoke) can damage the membrane lining the bronchi. That membrane contains cilia that sweep debris out of your lungs, but when the cilia are damaged, your body has difficulty cleaning out this debris. (For more on cilia, see Chapter 2.) So in an attempt to trap the debris and keep it from getting down further into your lungs, your body secretes more and more mucus, resulting in a *productive cough,* a cough in which you produce enough mucus to spit out. (You're spitting out the debris, too.)

Chronic bronchitis causes the small airways (bronchioles) to become less able to move air. (That is, airway resistance develops.) With less air moving through the airways, the body's oxygen/carbon dioxide exchange is affected. People with chronic bronchitis usually have a decreased oxygen level because they don't breathe in and out well or get oxygen to all their cells and tissues effectively.

Due to the restriction of oxygen, chronic bronchitis can harm other body systems besides just the respiratory system.

A low blood oxygen level can stimulate the kidneys to produce a hormone that causes the body to produce more red blood cells. Because the RBCs carry oxygen in their hemoglobin, the body attempts to improve the oxygen level by producing more RBCs. However, too many RBCs in the bloodstream is a condition called *polycythemia,* in which there is not enough hemoglobin present in the RBC for the oxygen to bind to. This causes the person to be oxygen-deprived despite the excess of RBCs. Looking at the lips and nails — or other places where blood vessels are near the surface, you can tell if a person isn't getting enough oxygen. If these are blue, the oxygen level may be low.

High blood pressure and heart problems also can occur in people with chronic bronchitis. Because of the constant attempts to increase the level of oxygen in the blood, the blood vessels in the lungs (pulmonary vessels) can spasm (tighten or constrict). Eventually, pressure in the heart's right ventricle increases, which can cause it to increase in size, resulting in possible heart failure and a condition called *cor pulmonale* (dilated right ventricle).

Cor pulmonale is the condition of the right ventricle of the heart being too large and dilated. Initially, enlargement of the right ventricle is an attempt after right-sided heart failure to get more deoxygenated blood into the lungs to become oxygenated. But with the lung problems caused by chronic bronchitis or other obstructive pulmonary disorders, cor pulmonale is difficult to manage and has a poor prognosis.

## Pneumonia

Pneumonia is an infection of the lung that can be fatal because it reduces the exchange of oxygen and carbon dioxide between the alveoli and capillaries. The infection, which can be caused by bacteria, a virus, or a fungus, results in mucus and pus building up in the alveolar sacs. The blockage in the alveolar sacs is what makes the oxygen/carbon dioxide exchange so difficult.

When the alveoli are blocked, the oxygen is unable to diffuse out of the alveoli into the capillary, and the carbon dioxide is unable to diffuse into the alveoli from the capillary. As a result, the person with pneumonia has a low level of oxygen in their bloodstream. *Bronchopneumonia* is a type of pneumonia that affects the deeper airways and alveoli. *Lobular pneumonia* affects part of a lobe in the lung; *lobar pneumonia* affects the whole lobe.

Bacteria, viruses, or fungi (collectively called *pathogens*) can get into the lungs through several routes. People can inhale pathogens, get them from contaminated medical equipment or if a person vomits and some of the vomit enters the trachea (that is, it's "breathed" in or *aspirated*), the contents of the stomach or throat can get into the respiratory system. The gastric acids can damage the mucosal lining of the airways, making them more susceptible to infection. Additionally, this solid gastric content can block airways and lead to an infection there that develops into pneumonia.

Regardless of what causes pneumonia, the symptoms are generally the same. Fever and shaking chills are common, as is cough and crackling lung sounds. Headache, joint pain, chest pain, sore throat, fatigue, shortness of breath, cyanosis (bluish color around mouth and fingernails), and increased heart rate also are experienced by some people.

People with normally functioning immune systems and strong respiratory systems can usually combat the pneumonia and restore their health to normal. However, several groups, such as smokers and alcoholics, are at higher risk of developing pneumonia and the complications of pneumonia along with those who suffer from

- ✔ Cancer

- ✔ Abdominal or thoracic surgery

- ✔ Chronic respiratory disease, such as asthma, cystic fibrosis, bronchitis

- ✔ Medications that suppress the immune system (such as corticosteroids)

- ✔ Malnutrition

- ✔ Sickle cell anemia

- ✔ AIDS

Treatment for pneumonia depends on the cause. Only bacterial infections can be treated with antibiotics; so viral pneumonia treatment is limited to caring for the patient's symptoms and hoping the patient's immune system is strong enough to battle the condition.

# Tuberculosis

Tuberculosis (TB) is a bacterial infection of the lungs. This bad-boy bacteria — *Mycobacterium tuberculosis* — is extremely difficult to eliminate. However, with early detection, treatment, although lengthy, usually is successful.

TB progresses as follows:

1. A person already infected with TB coughs or sneezes. If you inhale any of these droplets, the bacteria gets into *your* lungs.

2. Your immune system responds to the presence of bacteria like it does for any other infection: It sends white blood cells and macrophages to the site of infection (see Chapter 13) where the macrophages engulf the bacteria, and carry the bacteria to lymph nodes, just like they normally would.

3. However, the macrophages then start to act irregularly. They clump together, forming *tubercles*. Wherever the tubercles lodge, the surrounding tissue is killed, and scar tissue forms around the tubercle.

4. If enough tubercles form and become encapsulated by scar tissue, the lymph nodes become inflamed and can rupture. When they rupture, the infection is able to spread from your lymph nodes to surrounding tissue. Because lymph nodes exist all over your body, TB can easily spread from your lungs to other sites.

Tissues in the lungs also can become inflamed and die, and a hole in a bronchus or lung (such as from eroded tissue) can occur, which results in a condition called *pneumothorax* (air leaking out of the lung that causes the lung to collapse).

One of the problems with TB is that it takes so long for symptoms to occur that a person may be infected and be completely unaware of it. So the infected person can spread the bacteria by coughing or sneezing without realizing that others are being infected. There's a reason to encourage people to cover their mouth and nose when they sneeze or cough!

# Emphysema

Emphysema is a *chronic obstructive pulmonary disorder* (COPD). Commonly known as a smoker's disease, emphysema can also affect people who've had long-term exposure to chemicals, asbestos, or coal. Over time, exposure to these irritants damages the bronchioles. Eventually, they collapse, trapping air inside them. The trapped air can cause the tiny alveoli to rupture, thus eliminating some of the places where gas exchange take place. The elasticity of the lungs is decreased, making it hard for a person with emphysema to breathe.

People with this disease cough and have *dyspnea* (shortness of breath). As the disease progresses, and more and more alveoli are damaged, less and less oxygen gets into the bloodstream. This means that less and less oxygen is being sent to the brain. As a result of this lack of oxygen, people with emphysema often experience fatigue, irritability, and depression.

# Lung Cancer

The mother of all respiratory system diseases, lung cancer is extremely wide-spread because of the huge number of people who smoke. And lung cancer doesn't affect only men, as it once seemed. More and more women are being diagnosed with lung cancer, which directly correlates to the increase in women smokers in past decades. Now that women have been smoking for several decades, the effects are appearing.

Lung cancer follows a certain progression, similar to that of chronic bronchitis. (If you need a refresher on the anatomy mentioned in this list, refer to section "Knowing Your Respiratory Anatomy" earlier in this chapter.)

1. The cells in your respiratory membrane that line the bronchi thicken and harden. This happens in response to continual exposure to an irritant, such as cigarette smoke.

2. Cilia stiffen and are unable to sweep away debris. Particles of dirt and dust pollute your lungs.

3. Abnormal cells form in the membrane and clump together to form a tumor (a mass of tissue formed by the growth of the abnormal cells).

4. Some of the abnormal cells break away and lodge in another part of the lung, forming more tumors. This spreading action is *metastasis*.

5. As the tumors grow, the airways become blocked. Eventually the air supply is cut off, and your lung collapses.

6. After your lung collapses, the dirt and mucus that filled it become trapped, often resulting in infection such as pneumonia or an abscess (a collection of pus).

7. At this point, part or all of your lung needs to be removed.

Treatment for lung cancer depends on how early abnormal cells are detected. If the cancer is in an early stage, infections can be treated with antibiotics, the offending irritant can be avoided (such as quitting smoking), and cells may return to normal. Chemotherapy can be used to prevent or limit the spread of tumors. Surgery to remove part of the lung may be necessary if the cancer has progressed to a late stage.

# The Developing Human

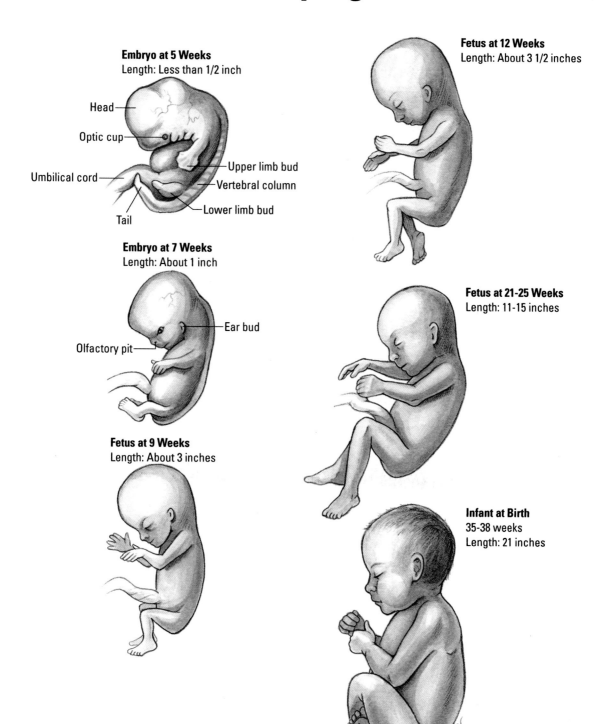

**Embryo at 5 Weeks**
Length: Less than 1/2 inch

Head
Optic cup
Umbilical cord
Upper limb bud
Vertebral column
Lower limb bud
Tail

**Embryo at 7 Weeks**
Length: About 1 inch

Ear bud
Olfactory pit

**Fetus at 9 Weeks**
Length: About 3 inches

**Fetus at 12 Weeks**
Length: About 3 1/2 inches

**Fetus at 21-25 Weeks**
Length: 11-15 inches

**Infant at Birth**
35-38 weeks
Length: 21 inches

K.Born

Note: 1cm = .4 inch

# The Nervous System

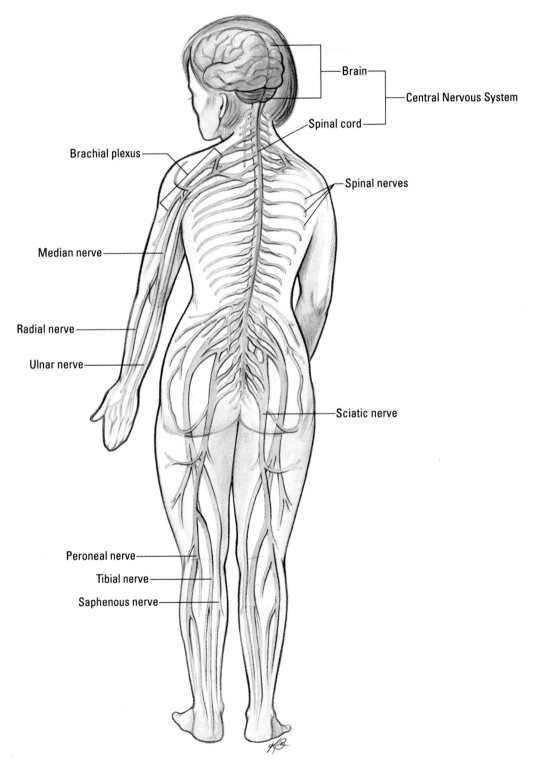

The central nervous system is comprised of the brain and the spinal cord. The peripheral nervous system includes 31 pairs of spinal nerves that connect the spinal cord to various parts of the body.

# The Glands of the Endocrine System

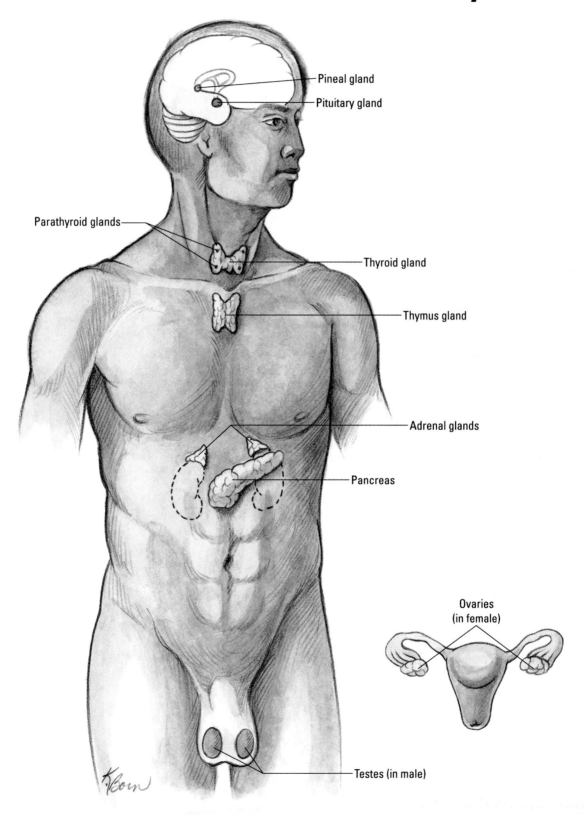

Pineal gland

Pituitary gland

Parathyroid glands

Thyroid gland

Thymus gland

Adrenal glands

Pancreas

Ovaries
(in female)

Testes (in male)

# Arterial Components of the Circulatory System

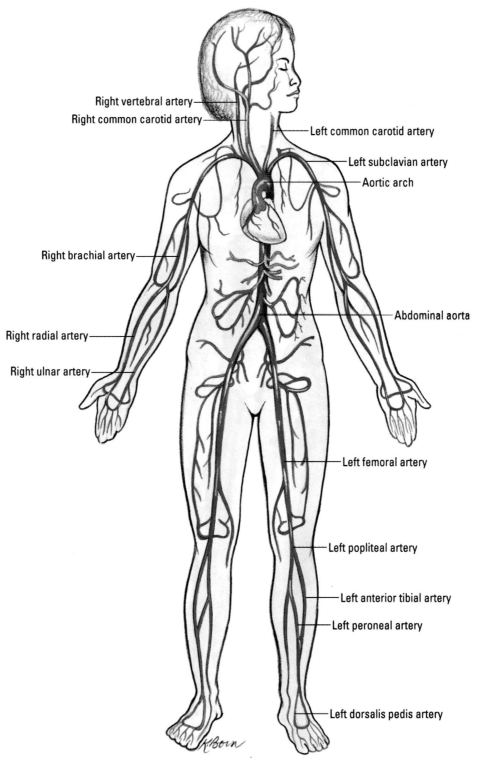

Right vertebral artery

Right common carotid artery

Left common carotid artery

Left subclavian artery

Aortic arch

Right brachial artery

Abdominal aorta

Right radial artery

Right ulnar artery

Left femoral artery

Left popliteal artery

Left anterior tibial artery

Left peroneal artery

Left dorsalis pedis artery

The arteries carry oxygenated blood from the heart to all parts of the body. Deoxygenated blood returns to the heart via the veins (not shown).

# Chambers of the Heart

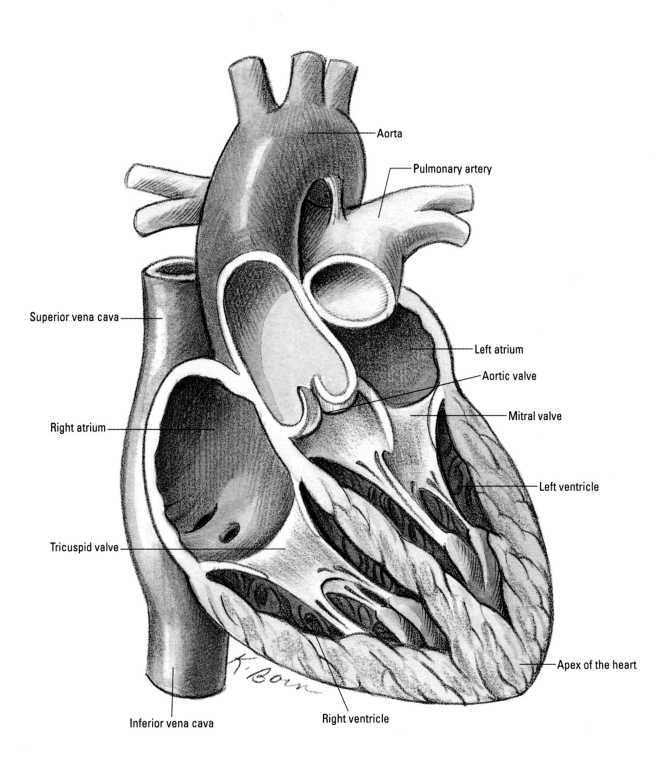

The four chambers of the heart, showing the valves and some of the major blood vessels.

# The Respiratory System

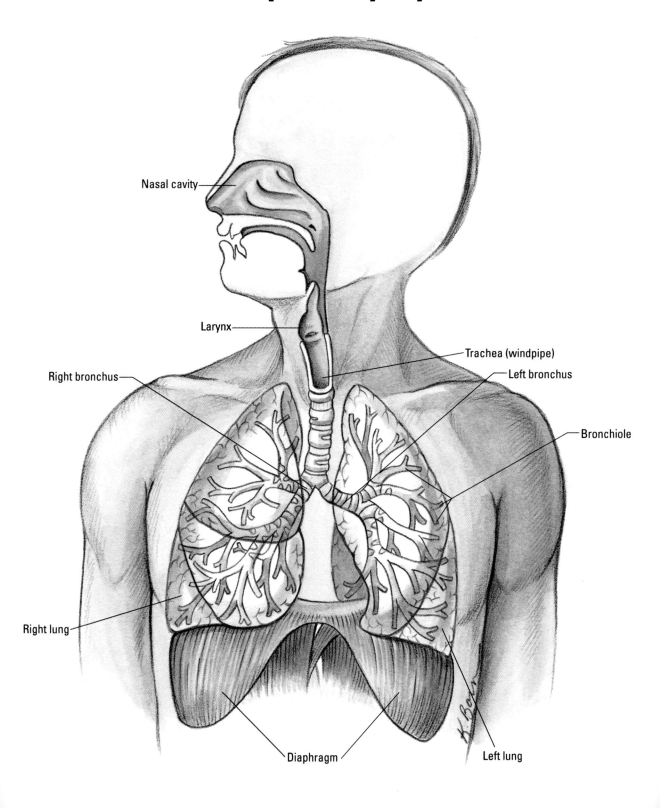

Nasal cavity

Larynx

Trachea (windpipe)

Right bronchus

Left bronchus

Bronchiole

Right lung

Diaphragm

Left lung

# The Respiratory System: Anatomy of the Bronchiole

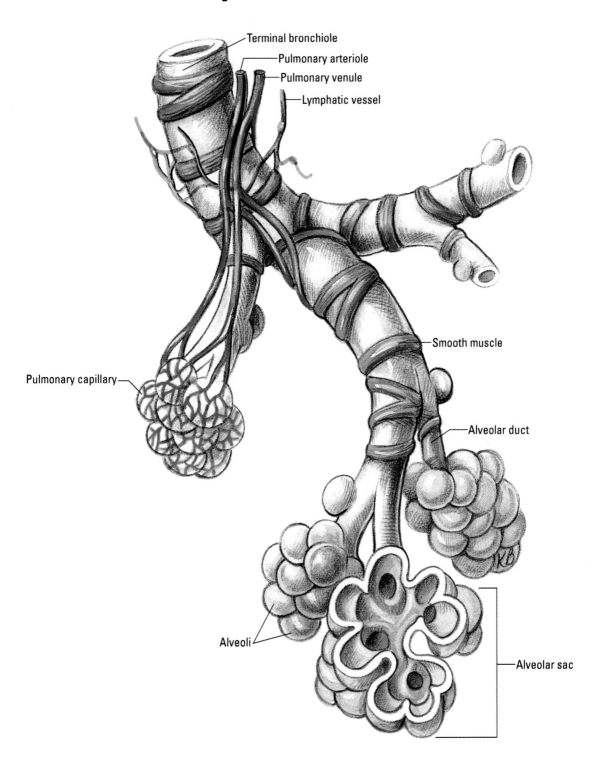

Terminal bronchiole

Pulmonary arteriole

Pulmonary venule

Lymphatic vessel

Pulmonary capillary

Smooth muscle

Alveolar duct

Alveoli

Alveolar sac

# The Digestive System

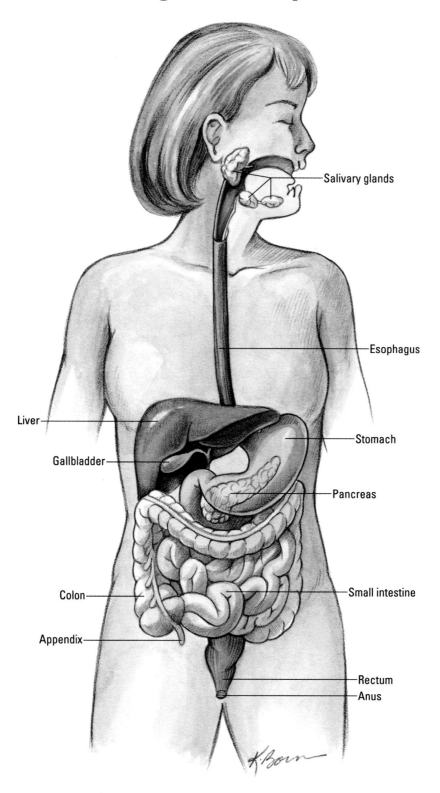

Salivary glands

Esophagus

Liver

Stomach

Gallbladder

Pancreas

Colon

Small intestine

Appendix

Rectum

Anus

K. Born

# Internal and External Features of the Stomach

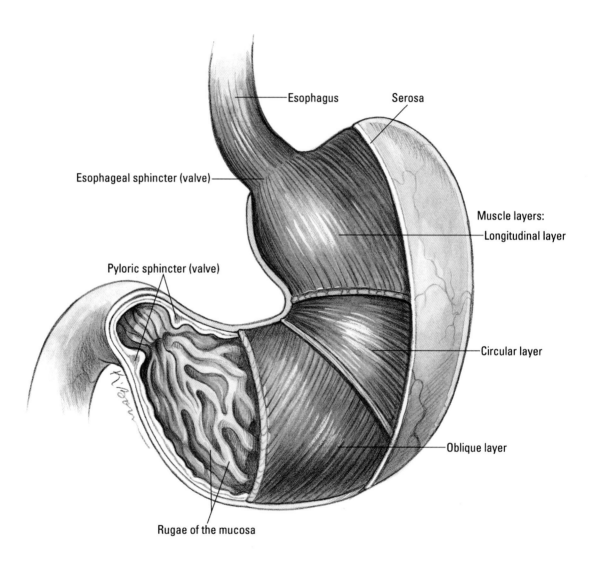

Esophagus

Serosa

Esophageal sphincter (valve)

Muscle layers:

Longitudinal layer

Pyloric sphincter (valve)

Circular layer

Oblique layer

Rugae of the mucosa

The mucosa secretes mucus, digestive enzymes, hydrochloric acid, and other substances that make up gastric juice.

# The Urinary System

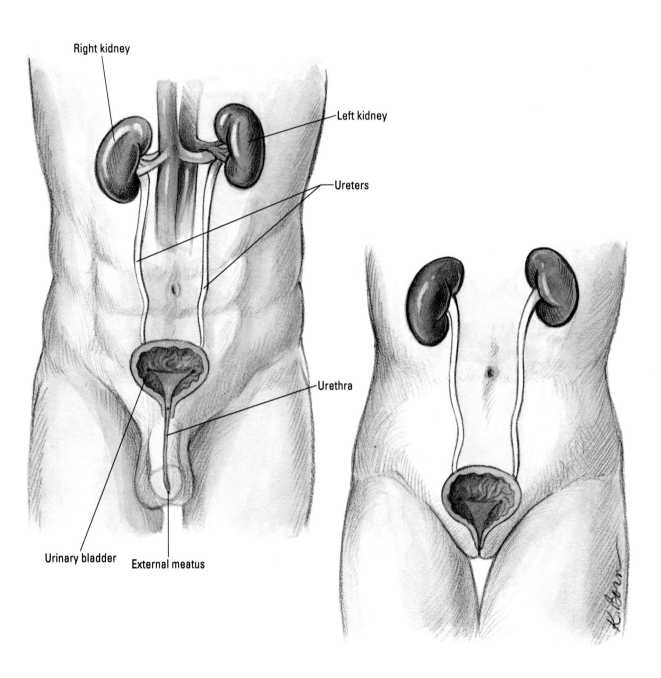

Right kidney

Left kidney

Ureters

Urethra

Urinary bladder  External meatus

# Blood Supply of the Nephron

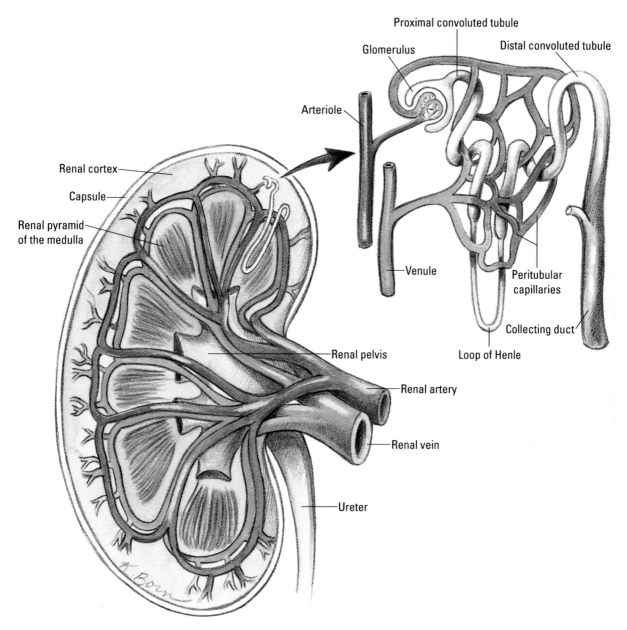

Proximal convoluted tubule

Glomerulus

Distal convoluted tubule

Arteriole

Renal cortex

Capsule

Renal pyramid
of the medulla

Venule

Peritubular
capillaries

Renal pelvis

Collecting duct

Renal artery

Loop of Henle

Renal vein

Ureter

K. Bonn

# Anatomy of the Kidney in Cross Section

# The Lymphatic System

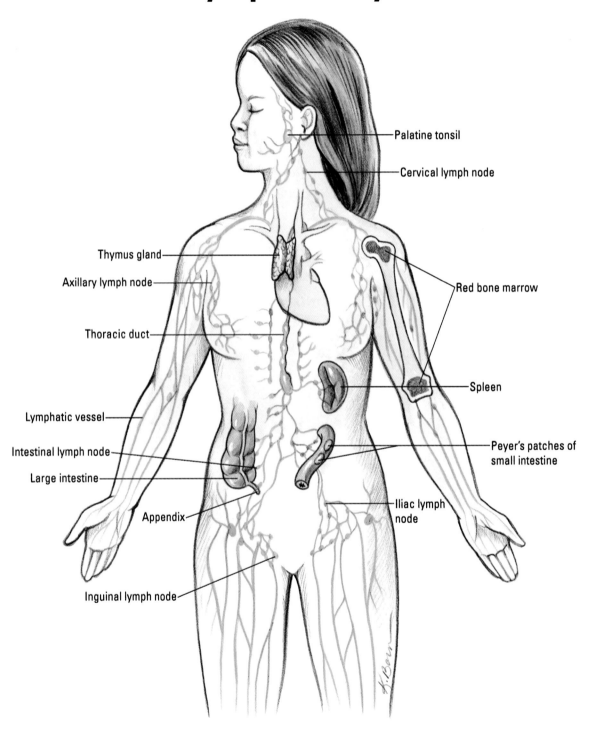

- Palatine tonsil
- Cervical lymph node
- Thymus gland
- Axillary lymph node
- Red bone marrow
- Thoracic duct
- Spleen
- Lymphatic vessel
- Intestinal lymph node
- Peyer's patches of small intestine
- Large intestine
- Appendix
- Iliac lymph node
- Inguinal lymph node

The lymphatic system consists of the lymphatic vessels and lymphoid tissue, such as found in the spleen, red bone marrow, and lymph nodes. Lymphoid tissue contains lymphocytes, which are cells that have immune function — they fight disease-causing organisms. The role that lymphocytes play in lymphoid tissue closely connects the lymphatic system with the immune system.

# The Female Reproductive System

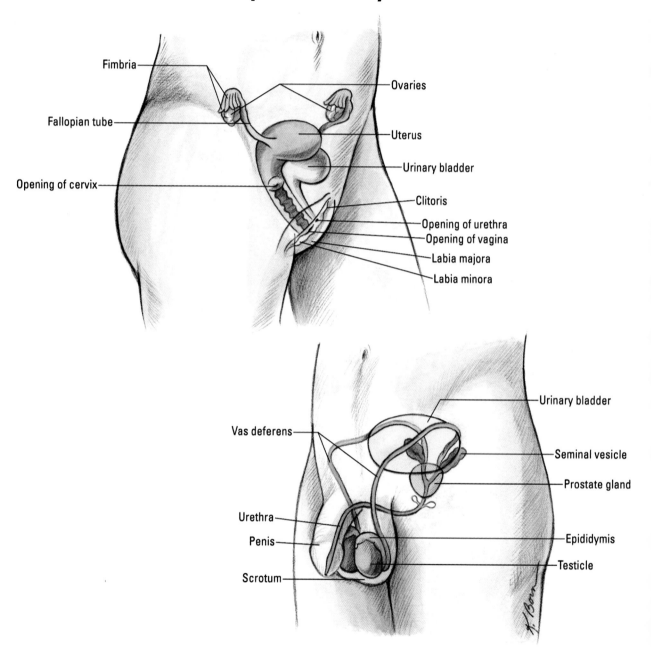

Fimbria

Fallopian tube

Opening of cervix

Ovaries

Uterus

Urinary bladder

Clitoris

Opening of urethra

Opening of vagina

Labia majora

Labia minora

Vas deferens

Urethra

Penis

Scrotum

Urinary bladder

Seminal vesicle

Prostate gland

Epididymis

Testicle

# The Male Reproductive System

# The Adult Skeleton

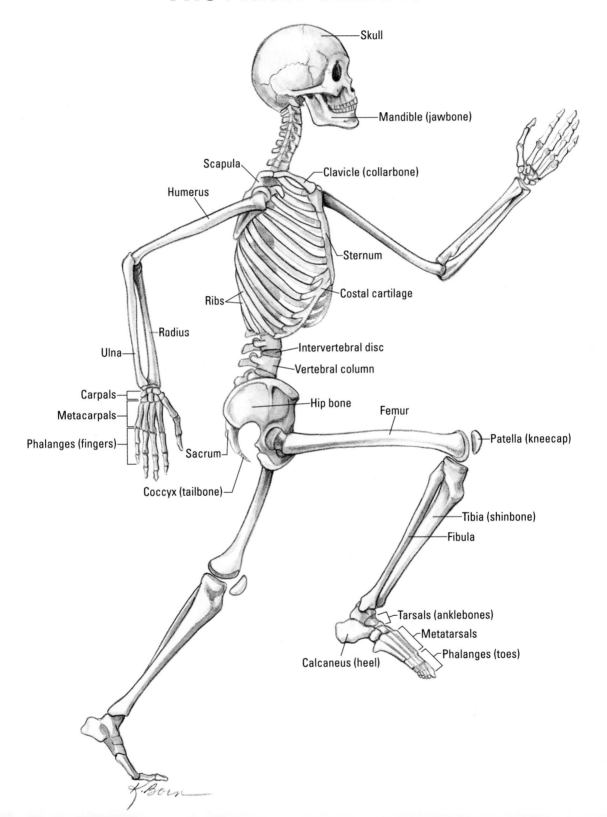

Skull

Mandible (jawbone)

Scapula

Clavicle (collarbone)

Humerus

Sternum

Costal cartilage

Ribs

Radius

Intervertebral disc

Vertebral column

Ulna

Carpals

Metacarpals

Hip bone

Femur

Phalanges (fingers)

Patella (kneecap)

Sacrum

Coccyx (tailbone)

Tibia (shinbone)

Fibula

Tarsals (anklebones)

Metatarsals

Phalanges (toes)

Calcaneus (heel)

K.Born

# The Muscular System

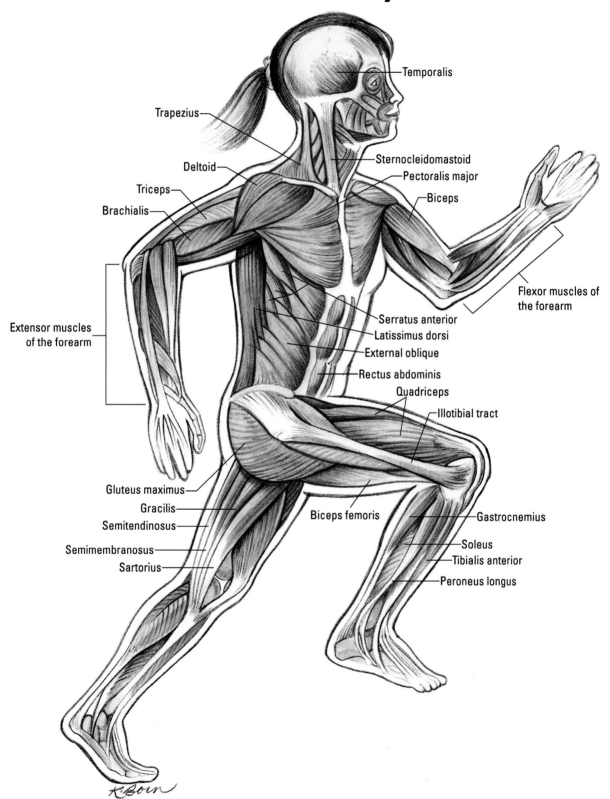

Temporalis

Trapezius

Deltoid

Triceps

Brachialis

Sternocleidomastoid

Pectoralis major

Biceps

Flexor muscles of the forearm

Extensor muscles of the forearm

Serratus anterior

Latissimus dorsi

External oblique

Rectus abdominis

Quadriceps

Illotibial tract

Gluteus maximus

Gracilis

Semitendinosus

Semimembranosus

Sartorius

Biceps femoris

Gastrocnemius

Soleus

Tibialis anterior

Peroneus longus

KBoin

# A Cross Section of the Skin

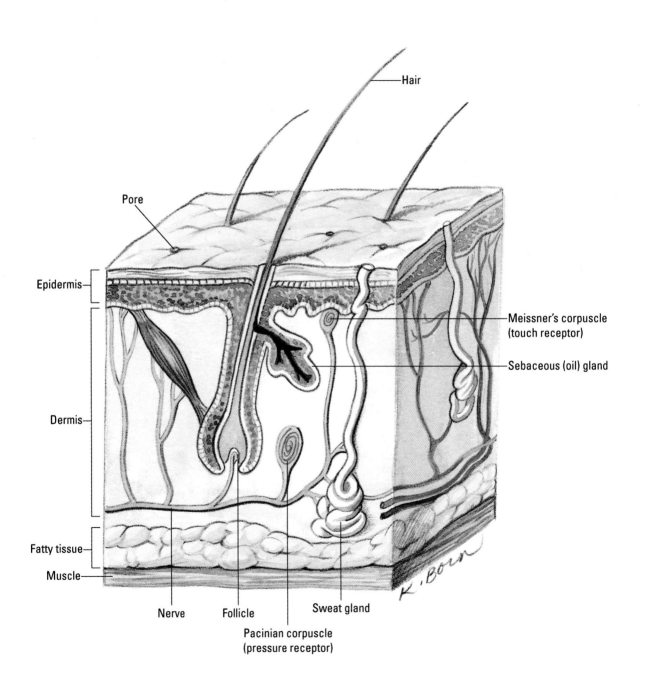

Hair

Pore

Epidermis

Meissner's corpuscle (touch receptor)

Sebaceous (oil) gland

Dermis

Fatty tissue

Muscle

Nerve

Follicle

Sweat gland

Pacinian corpuscle (pressure receptor)

K. Born

# Chapter 11

# Breaking Down: The Digestive System

*L*ike the respiratory system, the digestive system is closely connected to the bloodstream. It has to be. Every cell in your body needs nutrients in order to survive. The nutrients come from the food, so after the digestive system breaks down the food, it must shuttle the nutrients into the bloodstream to supply the cells with what they need. This chapter explains this process as well as the digestive system's anatomy (see Figure 11-1) and how waste is removed from your body. I also explain what nutrients your body needs, so you can make better decisions about the foods that you choose to eat. At the end of the chapter, you can find some of the causes and symptoms of digestive pathophysiology along with common treatments.

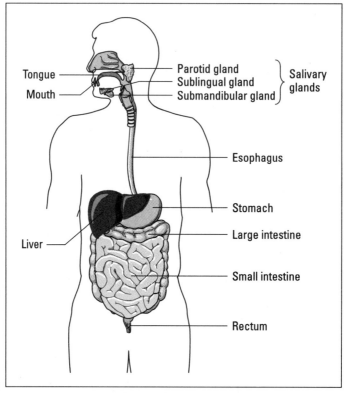

**Figure 11-1:**
The organs
of the
digestive
system.

# Following the Path of Food

The digestive system, not surprisingly, is responsible for digesting (breaking down) food. But the digestive system has other functions, too. It ingests food, absorbs the small nutritious molecules from broken down food, and eliminates food particles that your body can't digest.

The digestive system's ability to break food down into the nutrients that your cells can use is what makes this system so important. Because every cell that makes up every tissue that makes up every organ needs nutrients to continue living — all your organs depend on your body's digestive organs to supply them with the fuel they need to keep on keepin' on.

# Now that's a mouthful!

Your mouth *(oral cavity)* is the starting point of your digestive system, a gateway to your other organs of digestion. Besides making eating a fun experience, your mouth serves some important digestive functions.

## Talking about your teeth and gums

Your teeth tear and grind food into small enough parts to digest easily. Humans have 32 teeth that are of four basic types:

- **Canines** tear food like meat.
- **Incisors** bite (think scissors).
- **Premolars and molars** are flat for grinding food.

The gums hold teeth in position, and the roots of your teeth are embedded in your jawbone by a binding material called *cementum.* Blood vessels that run through the jaw bone and up into the pulp of the tooth supply the teeth with blood. *Dentin,* a bonelike material, covers the pulp, and an extremely hard protective enamel covers the dentin.

## Sinking your teeth into it

If you don't brush and floss your teeth after eating, you can get cavities. If you don't remove the bits of food off your teeth, bacteria, which are always present in your mouth, break them down for you. In the process, acids are produced, and those acids eat away the enamel, creating a cavity. If a cavity gets deep enough, bacteria can make their way into the pulp of the tooth. Besides causing great pain, an infection called an *abscess* can result.

## Holding your tongue

Your tongue is made up of muscle tissue just like the muscle found in your arms and legs. However, it's covered with a mucous membrane as well as taste buds that detect the chemicals in foods and send signals to your brain via attached nerves (see Chapter 7). I don't mean just the chemicals like monosodium glutamate, Red Dye #40, or Blue 1. Those are additives and artificial colors. I am talking about the chemical elements that make up even the most natural of foods.

An apple contains natural sugar (fructose) and other carbohydrates (starches). The taste buds that are sensitive to sweet detect the sugars and send a message to your brain, which then figures out what you're eating and creates the sense of taste. Raw broccoli contains certain chemicals that give it a bitter taste. The taste buds on your tongue that detect bitter tastes send a signal to your brain, which most likely decides that you need some dip to cover up the bitter taste.

Muscles attach your tongue to your skull bone, and a mucous membrane on the underside of the tongue attaches the tongue to the floor of the oral cavity. That stringy piece of membrane that you see when you touch the tip of your tongue to the roof of your mouth is the *lingual frenulum*.

### The spitting image of saliva

Also inside the mouth, several *salivary glands* are the first glands to secrete substances during digestion — often before you even eat anything. Just smelling food or the anticipation of eating something you enjoy can "get the juices flowing." The three types of salivary glands are

- **Sublingual glands:** Under the tongue.
- **Submandibular glands:** Located below the mandible but inside the oral cavity, putting them below the tongue also.
- **Parotid glands:** Just inside the cheek right below your ears. These glands swell and thus create "chipmunk cheeks" during the viral infection called *mumps*.

Saliva contains mucus (to moisten food and aid swallowing) and the enzyme *salivary amylase,* which immediately begins breaking down carbohydrates (starches) in foods as soon as you put them into your mouth. Enzymes are proteins that speed up chemical reactions. (For more on the many enzymes in the digestive system, see the "Stirring it up in your stomach" and "Doing the Chemical Breakdown" sections.)

## No, the pharynx and esophagus are not Egyptian landmarks

The *pharynx,* better known as your throat, marks the beginning of the *esophagus,* the tube through which food and drink passes to get from the mouth and throat into the stomach. As you chew your food into little pieces, your tongue moves the pieces around and, along with the saliva, forms a ball called a *bolus.* When you swallow, the bolus bounces off a piece of cartilage called the *epiglottis* and so is diverted from the larynx and falls into the esophagus.

"Through the lips and over the gums. Look out stomach, here it comes!" You may have heard that old toast of good cheer before, but I think it needs to be changed. How's this? "Through the lips and down the esophagus. Look out stomach, I'm gonna swallow this!" Much more accurate, I think. See, the old version was missing the link between the mouth and the stomach, as well as the way that the ingested or imbibed substance was going to get into the esophagus. Now, I've covered it. Be sure to use it at your next party.

The esophagus has a sphincter muscle at the top and a sphincter muscle at the bottom through which the bolus passes. The sphincter at the top is the *pharyngoesophageal sphincter.* A muscular action called *peristalsis* is what moves the bolus down through the esophagus. The lining of the digestive tract contains muscular tissue, and as it contracts, it squeezes the bolus through the tube. Picture yourself squeezing toothpaste out of a tube. Your hand squeezing the tube is similar to the action of peristalsis, and the bit of toothpaste that is forced out is like the bolus. When peristalsis has moved the bolus to the bottom of the esophagus, the *gastroesophageal sphincter* opens to allow the bolus to enter the stomach.

## Stirring it up in your stomach

When you eat, a bolus of food plops into your stomach, but you don't feel the plop. The hollow organ known as the stomach is made up of four different layers:

- ✔ **Serous coat:** Covers most of the stomach
- ✔ **Muscular coat:** Forms the hollow and churns food
- ✔ **Submucosal coat:** Connects the muscular and mucosal layers
- ✔ **Mucous coat:** Protects the inner lining of the stomach

The muscular coat has three layers of muscle fibers, each of which run in different directions. The three different types of muscle fibers allow the stomach to contract as necessary to churn the food you've ingested.

You feel full when the stomach's *stretch receptors* reach a point where they touch off a nerve signal to the brain and your brain senses that you're full. (See Chapter 7 for more on how nerves send signals to the brain.) The stretch receptors are located in the wall of your stomach.

The lining of the stomach contains ridges called *rugae,* which provide more surface area. As the stomach fills, the rugae smooth out, allowing the stomach to expand. Picture the sheets on your bed all messed up and lumpy. The "lumps" are like the rugae. When the sheets are lumped into a ball, they don't cover your entire bed. But, when you smooth out the lumps, the surface area increases, covering the mattress.

Just like the esophagus and the mouth (feel the inside of your cheek), a mucous membrane covers the inside of the stomach. Actually, the mucous membrane runs continuously from the mouth all the way through the entire digestive system. One function of the mucous membrane is to protect your digestive organs from being eaten away by the strong acids secreted in the digestive system, such as hydrochloric (gastric) acid.

*Gastric juice* is secreted from the millions of tiny gastric glands that are part of the mucosal lining of the digestive system. It contains *hydrochloric acid* (HCl) and *pepsinogen* (the precursor to the enzyme *pepsin*), which starts breaking down proteins into *peptides*. HCl is one of the most acidic substances found anywhere. The advantage to it being so strong is that it kills bacteria present in food, it converts pepsinogen into pepsin, and it helps to break down the connective tissue in meats, in case you ever wondered how you digested that piece of steak you swallowed.

The hydrochloric acid in gastric juice also stimulates the cells lining the stomach to secrete *gastrin* (a hormone). The gastrin is absorbed through the lining of the stomach into your capillaries then travels through your bloodstream. The gastrin regulates the release of HCl, mucus, and pepsinogen that make up gastric juice. As long as gastrin is flowing in your bloodstream, your stomach continues to secrete gastric juice. That means that as long as you keep eating, your stomach keeps secreting the acids that start breaking down your food. This loop ensures that there is enough gastric acid available for the amount of food you need to digest. (For more on gastrin, gastric juice, and the stomach, see Chapter 8.)

The stomach's action of contracting to churn up its contents is part of physical digestion, like chewing, swallowing, and peristalsis. But it's the stomach's contribution to chemical digestion that really helps break down the food you eat. (See the "Doing the Chemical Breakdown" section later in this chapter for more on chemical digestion.)

After the stomach churns and gastric juice mixes with the bolus of partially digested food, the food breaks down even more completely and turns into an oatmeal-like paste called *chyme*. This process takes anywhere from two to six hours. The chyme then squirts into the small intestine through the *pyloric sphincter,* a muscular ring between the lower part of the stomach — called the *pylorus* — and the top of the small intestine — called the *duodenum.*

To envision chyme squirting into your small intestine, picture thick icing squirting out of a pastry bag, the bag being your stomach, the tip of the bag being your pylorus, and the thick icing being your chyme. Yum!

## *Testing your intestinal fortitude*

The duodenum (the top of your small intestine), which receives chyme from the stomach, is one foot in length, but it's just the beginning of chyme's long journey through your body. The next section of the small intestine, the *jejunum,* is about three feet long. The last section, the *ileum,* is about seven feet long. Eleven feet of digestive tubing doesn't seem to make the "small" intestine small; but the intestines are not classified as small and large based on their length. Rather, the width of the intestines is the characteristic that differentiates the two. The small intestine is much longer but more narrow than the large intestine.

Both the small and large intestines are made up of two layers of muscle tissue — longitudinal and circular — and lined with mucosal tissue.

### Doing most of the digesting: The small intestine

The small intestine does the lion's share of the work in the digestive system. When the chyme arrives in the duodenum, *Brunner's glands* in the walls of the small intestine secrete mucus and bicarbonate to help neutralize the strong acids in the chyme that just came from the stomach.

The cells in the walls of the small intestine secrete the hormones *secretin* and *cholecystokinin* (CCK), making the small intestine a gland as well as a digestive organ. When the small intestine secretes these two hormones, they're absorbed into the bloodstream and carried to their target organs — the gallbladder and pancreas — where they stimulate the gallbladder to release bile and the pancreas to secrete pancreatic juice. (See the "Doing the Chemical Breakdown" section later in this chapter.)

The small intestine does get some help from other digestive organs.

In addition to mucus, the cells in the walls of the duodenum also secrete digestive enzymes. The cells from which the enzymes are secreted reside in a place with a really cool name: the *crypts of Lieberkühn*. It sounds like the name of a fantasy, action video game, but they're really microscopic pits in the walls of the small intestine. The crypts secrete the enzymes that pass into the duodenum through ducts from the pancreas and liver.

By the time the chyme leaves the duodenum, it's almost totally digested. The carbohydrates, proteins, and fats that made up the food that you originally put in your mouth several hours before have been broken down into molecules that the bloodstream can absorb, such as glucose, amino acids, fatty acids, and glycerol, respectively. After these molecules are available, peristalsis — the muscular contractions that push digested food along — moves it into the jejunum and ileum of the small intestine.

Throughout the combined ten feet of jejunum and ileum, several million villi and microvilli stretch along the mucosal lining. *Villi* are fingerlike projections of mucosal lining that reach into the hollow space of the small intestine. A capillary that connects to each villi is capable of directly absorbing nutrients into the bloodstream where they become available to every nook and cranny of the body via the circulatory system. (See Chapter 9 for more on the circulatory system.)

*Microvilli* are even smaller projections off of the villi that serve as a point of absorption of nutrients. The nutrients — such as glucose, amino acids, or needed electrolytes — are carried through the microvilli into the cells of the villus. Because the nutrients are "carried," the process requires active transport. And active transport requires the expenditure of some energy in the form of ATP. Active transport occurs again to carry the nutrients out of the villus and into a capillary. (For more on active transport and ATP, see Chapter 3.)

The two products of fat digestion — fatty acids and glycerol — are handled a bit differently from one another. Short-chain fatty acids are shuttled directly to the capillary through the microvilli and villi. But the long-chain fatty acids are transported from the villi through the lymphatic system. The body's cells put the long-chain fatty acids back together as compounds called *triglycerides*. Running up inside each villi, a lymphatic vessel called a *lacteal* absorbs triglycerides.

By the time the food you ingested works its way through all three parts of your small intestine, the nutrients that your body needs have already been absorbed into the bloodstream, and all that's left at the bottom of the small intestine — the ileum — is the junk that the body doesn't want. This leftover material gets passed onto the large intestine, which has the job of disposing of the leftovers.

### It's a crappy job, but your intestines have to do it

Okay. Back to the small intestine for a second. When all the food has been digested to the smallest possible pieces, the pasty chyme oozes from the small intestine to the large intestine (also called the *colon*). No sphincter involved here. No enzymes breaking down carbohydrates, proteins, or fats. No churning of contents. As chyme is ready, it simply passes through the last portion of the small intestine — the ileum — through the *ileocecal valve* into the *cecum,* the first portion of the large intestine.

The large intestine, which is about six feet long, frames the small intestine. Beyond the cecum, the large intestine moves upward as the *ascending colon,* across as the *transverse colon,* and downward as the *descending colon.* (Nobody said anatomists were creative; if they were, the names of body parts would be much harder to remember.) The *rectum* holds the feces (poop) until you have a chance to release them; the process of releasing feces is *defecation.* As defecation is taking place, the feces pass through the *anal canal* and exit the body through the *anus.*

The major event that happens in the large intestine is that water from the chyme is absorbed through the lining of the large intestine. The water diffuses into the capillary and goes back into the bloodstream. The removal of water from chyme compacts the indigestible material in the colon, forming *feces.* In addition to undigested food, the feces contain dead cells from the body. The brown color of feces comes from the combination of greenish-yellow bile pigments, broken down hemoglobin, and bacteria. The bacteria that's most prevalent in the large intestine is *Escherichia coli (E. coli).*

The *E. coli* bacteria eat some of the indigestible material in your feces (and you think Brussels sprouts are bad!); as they do, they produce molecules of digestion that have a well-known odor. I know it, and you know it. It's nothing to be embarrassed about (or proud of). Still, *E. coli* and other bacteria in your colon are beneficial to you in that they finish digesting some more material in your digestive system (but apparently they don't like corn or peanuts).

Some of these beneficial bacteria actually produce vitamins that your body needs in the colon. For instance, *vitamin K,* which is necessary for proper blood clotting, is produced in the large intestine by bacteria, absorbed through the intestinal lining, and transported directly into the bloodstream via the capillaries.

As the colon completes its work, contractions move feces into the rectum, which is located at the bottom of the colon. When the rectum has about 150 to 200 grams of feces in it, it stretches. Stretch receptors in the rectum send a signal to the brain through a nerve that register as the need to defecate. Normally, your anal sphincter is contracted to keep feces from exiting your anus at inappropriate times. When you are ready to defecate, you relax your anal sphincter, and it opens, allowing the feces to come out.

# Doing the Chemical Breakdown

When you hear that the digestive system is responsible for breaking down food, you probably think that organs of the digestive system perform those tasks. Indirectly, they do. But most of the organs of the digestive system work as glands — they secrete the hormones (and enzymes) that *really* do the dirty work.

The pancreas, liver, and gallbladder are often referred to as the *accessory organs of digestion* because they don't directly churn, break down, or squeeze foodstuffs through the digestive system. But although they aren't involved in physical digestion, they're involved in the chemical digestion processes.

## *Pigging out with your pancreas*

The pancreas sits next to the duodenum of the small intestine and behind the stomach. This gland is known for secreting insulin, which helps the body control the blood's glucose level. The secretion of insulin is an endocrine function of the pancreas. (For more on the pancreas and insulin, see Chapter 8.) However, the pancreas also secretes a number of other substances that are important to digestive functions. *Pancreatic juice* contains enzymes that digest carbohydrates, fats, and proteins. It also contains sodium bicarbonate (the same compound as in baking soda), which plays an important role in reducing the acid in the chyme from the stomach. (For more about chyme, refer to the "Stirring it up in your stomach" section earlier in this chapter.)

Reducing the acidity is necessary because the enzymes are effective only when the mixture of chyme and juices is neutral. If the mixture is too acidic, the enzymes don't work, and digestion can be incomplete, leading to intestinal trouble. Nearly every cell of the pancreas secretes pancreatic juice and passes it through the pancreatic duct directly into the duodenum of the small intestine.

Some of the enzymes that are found in pancreatic juice include trypsin, peptidase, nuclease, lipase, pancreatic amylase, lactase, sucrase, and maltase. Table 11-1 shows you what each pancreatic enzyme does.

| Table 11-1 | Functions of Pancreatic Enzymes | |
|---|---|---|
| *Enzyme* | *Targeted Nutrient* | *Result of breakdown* |
| Trypsin | Proteins | Peptides (chains of amino acids) |
| Peptidase | Peptides | Individual amino acids |
| Lipase | Fats | Fatty acids and glycerol |
| Nuclease | Nucleic acids (DNA, RNA) | Nucleotides |
| Amylase | Carbohydrates | Glucose and fructose |
| Sucrase | Sucrose | Glucose and fructose |
| Maltase | Maltose | Two glucose molecules |
| Lactase | Lactose | Glucose and galactose |

## Filtering through your liver

The liver, the body's largest gland, filters blood through its approximately 100,000 tiny lobules. These lobules form the two lobes of the liver. Each lobule is filled with bile canals and has access to the hepatic portal vein and hepatic artery. (*Hepatic* refers to the liver.)

The liver's filtering ability is closely related to its structure and location in the body. Your liver sits right under your diaphragm and right above your stomach on the right side of your abdomen. It basically is right in the center of your body where it can keep a check on your blood.

It filters toxins (such as alcohol and drugs) out of the blood that flows from the intestines through the hepatic portal vein to the liver, and it helps to maintain constant levels of important blood components. The liver also produces bile, which the bile ducts transport to the gallbladder. (For more on bile, see the "Floating down the Bile" section.)

The liver produces several substances and has other functions in addition to filtering, creating bile, and breaking down red blood cells.

- ✔ After you eat a large meal, your liver stores excess glucose as *glycogen;* then breaks it down between meals to keep your blood glucose level steady — a natural time-release process.

- ✔ The liver produces *urea* after amino acids are broken down during digestion and removal of dead cells.

- ✔ After the liver removes dying red blood cells from the bloodstream, the hemoglobin is converted to the bile pigments, and the liver recycles the iron held in the heme molecule.

- ✔ The liver produces *plasma proteins,* such as fibrinogen, that travel in the bloodstream to help repair damaged vessels and tissues.

## Gliding on through your gallbladder

Your gallbladder is a greenish little organ tucked into the curve of your liver that serves to store excess bile. As needed, the bile is released from the gallbladder and passes out of the gallbladder through the *cystic duct* in the "neck" of the gallbladder. Then the bile flows right through the *common bile duct* into the duodenum of the small intestine, near where the pancreatic duct enters the duodenum. In the duodenum, the bile mixes with the chyme coming out of the stomach and flowing into the small intestine.

## Floating down the Bile

Bile emulsifies fats. During the *emulsification* process, lipase (one of the enzymes in the pancreatic juice) breaks down large fat molecules into little droplets called *micelles.* You know how oil and water don't mix? Well, emulsifiers, such as bile, are substances that help them to mix. Emulsification liquefies fats so that they can be carried through the fluid bloodstream without solidifying just like emulsifiers are added to foods to keep fats and oils from solidifying.

As the fats in foods you previously digested are broken down, the micelles surround them and carry the fats to the microvilli in the lining of the small intestine where they're released into the lacteals. (See the "Doing most of the digesting: The small intestine" section.) Another purpose of bile is to help absorb the fat-soluble vitamins K, D, and A into the bloodstream more easily.

# Digestive System Diseases and Disorders

Yes, just like every other system in the body, your digestive system can go out of wack. If you've had an intestinal flu or a stomach virus, you probably experienced two of the most common human maladies: vomiting and diarrhea. However, many problems can crop up in the digestive system, and their severity ranges widely. This section gives you some info on several of the most common diseases and disorders of the digestive system.

## Worming its way into your appendix: Appendicitis

Your appendix, also called the *vermiform* (meaning "wormlike") *appendix,* is indeed a structure that looks like a worm. A little sac attached to the cecum at the beginning of the large intestine, the appendix doesn't really have a true function anymore; it's considered a vestigial structure, meaning that it's just hanging around doing nothing now that evolution has passed it by.

During the transfer of chyme from the small intestine to the large intestine, some material may get into the appendix. Normally, the material that gets into the appendix does make its way out. But, if it doesn't come out, depending on the material deposited into the appendix or how long it remains filled, the appendix can become inflamed or infected, causing *appendicitis.*

Appendicitis can occur suddenly (an acute attack) and is considered to be a medical emergency. Initially, the pain of appendicitis is in the upper right side of the abdomen. But as the inflammation worsens, the pain spreads to the lower right side. A person with appendicitis may be nauseous, and the body's attempt to kill off the bacteria infecting it may cause a fever.

The appendix can become blocked as the tissue swells from inflammation. When the appendix is blocked, the mucus released in its lining cannot leave the structure, and the appendix swells. The pressure builds inside the appendix, and the muscle tissue of the appendix contracts in an attempt to force the contents out, causing great pain. If the contents aren't expelled, bacteria continue to multiply, and the infection and inflammation worsens. The pressure can build to such a level that the appendix ruptures, spilling the contents and bacteria into the abdomen. If the appendix bursts, the situation can be life threatening and requires emergency surgery.

## Lacking water: Constipation

Constipation results when the large intestine absorbs too much water out of the feces, which makes the feces dry, hard, and a bit painful on exit. The most common cause of constipation is ignoring your body's signal to defecate. The feces remain in the colon too long, and too much water is absorbed into the lining of the intestine, drying out the feces.

Another cause is too little fiber in the diet. Fiber gives your large intestine a workout, keeping the muscles able to contract well enough to force out the feces. If they don't work at their optimal level, fecal material remains in the large intestine. This situation can not only result in constipation, but also colon cancer if it happens regularly.

If fecal material isn't totally evacuated from the large intestine, or if it remains in the colon too long (such as with constipation), the cells of the intestinal lining can be damaged, infected, scarred, and then less effective. Increasing fiber in the diet can strengthen the intestine, move feces through the large intestine better, and more completely clear out fecal material.

Dietary fiber, such as from whole grains, vegetables, and fruits, or bulk laxatives (those with fiber — *not* chemicals) are the "prescription" for constipation. And, oh yeah, don't forget to drink more water, too. Water helps both constipation and diarrhea.

## Cranky Crohn's disease

*Crohn's disease* is an inflammatory bowel disease, meaning that the lining of the intestine becomes inflamed. The mucosal layer, the muscular layer, the *serosa* (the covering tissue), and even the lymph nodes and membranes that provide blood supply to the bowel can be affected. As the intestinal lining swells, ulcers, fissures (cracks), and abscesses (pus-filled pockets) can form.

Crohn's disease can affect not only the large intestine, but the small intestine, too. Most often, the ileum, the final part of the small intestine, is affected. In the early stages of the disease, people have diarrhea and do not feel well. They may also experience a fever, as well as pain in the lower right side of the abdomen.

As Crohn's disease progresses, areas of lymph follicles, called *Peyer's patches,* form on the lining of the small intestine. Those follicles make the intestinal wall thicken and stiffen as fibers are deposited, which makes the hollow space inside the intestine narrower. Diarrhea becomes severe because the water from feces is unable to be absorbed through the diseased lining. As segments of bowel become more diseased, they begin to stick to other diseased segments, which shortens the bowel. Complications such as *fistulas* (abnormal openings between the skin and an organ or between two organs) or obstructions can occur. Patients with Crohn's disease often have an anal fistula (an opening between the diseased bowel and the skin near the anus) or between the bowel and the vagina or bladder.

Crohn's disease sufferers develop nutritional deficiencies because of poor absorption and problems with digestion along the route in the digestive system. Usually, people with Crohn's disease become deficient in vitamin $B_{12}$ because the severe diarrhea flushes everything from soup to nuts (in their digested forms, of course) to the beneficial bacteria that live in the intestines right out of the digestive system. This deficiency can lead to a condition called pernicious anemia.

Treatment for Crohn's disease includes dietary changes, rest, stress reduction, vitamin supplements, and medications to reduce inflammation and pain. Surgery sometimes is necessary, including removal of the large intestine and ileum of the small intestine.

## Doing the diarrhea trot

Rather than being absorbed by the large intestine and put back into the bloodstream, an excessive amount of water remaining in the feces causes diarrhea. Because water is leaving your body, it's easy to become dehydrated when you have diarrhea. Some people may think that you shouldn't drink liquids when you have diarrhea because taking in more fluid makes the diarrhea worse. Nope. Sorry, it doesn't work like that. In fact, it's totally the opposite, so drink up.

Bacteria infecting the large intestine (that is, bacteria that don't normally inhabit the large intestine) can also cause diarrhea; the bacteria get into the large intestine usually through contaminated food (food poisoning). When the lining of the intestinal tract is irritated or inflamed, peristalsis — the muscular contractions that move materials through the digestive system — increases in an attempt to get the bacteria out of the system fast. With feces moving through the large intestine faster, the water doesn't have enough time to be absorbed back into the body. The water is lost through defecation. So when bad food causes you to have diarrhea, let nature take its course. Drink plenty of water to compensate for the losses and flush out the bacteria. Anti-diarrhea medicines and electrolyte replacement solutions are not necessary at the first sign of diarrhea. If diarrhea continues with no sign of letting up, then call your doctor.

Another cause of diarrhea is stress. When you're nervous, worried, tense, anxious, or otherwise on pins and needles, your nervous system is in hyper-alert mode. You may recall that the *flight-or-fight response* initially caused stress in humans: Man sees saber-toothed tiger; tiger pounces; man secretes adrenaline to make his hairy legs run faster. That kind of stress warranted a bit of nervousness in order to preserve life. But saber-toothed tigers are now extinct, so lighten up! The stress of everyday life — having to be in more than one place at the same time; traffic tie-ups; long lines; work pressures; financial, family, or relationship problems — all of that does indeed draw a response from your nervous system. And, sometimes that response can stimulate the release of adrenaline, which also speeds up peristalsis. Just like the presence of bacteria in the large intestine, nervousness can move feces out of the large intestine too fast, keeping water from being reabsorbed and causing diarrhea.

## *You've got gall: Gallstones*

If the liver has made too much cholesterol or the bile becomes concentrated, cholesterol crystallizes. If the crystals grow, they become *gallstones.* Gallstones can fill up the gallbladder, reducing the amount of space for bile to be stored. Or, worse, the gallstones can block the common bile duct leading to the small intestine, which can cause a condition called *obstructive jaundice.* A buildup of bile pigments in the blood cause *jaundice,* a noticeable yellowing of the skin. If the common bile duct is blocked, the bile pigments do back up into the blood. A laparoscopic laser technique called *lithotripsy* can obliterate the gallstones without major abdominal surgery, which carries high risks for infection.

# Honing in on hepatitis

An inflammation of the liver is called *hepatitis*. Usually, a virus causes hepatitis; both the hepatitis and the virus can be one of five types:

- ✔ **Hepatitis A:** Most commonly spread through water contaminated with feces, which is why restaurant employees must wash their hands after going to the bathroom.

- ✔ **Hepatitis B:** Most commonly spread through sexual contact, this type affects many people with human immunodeficiency virus (HIV). Contaminated needles or blood transfusions can also spread this hepatitis. A vaccine is available.

- ✔ **Hepatitis C:** Contaminated blood spreads this type of hepatitis, but no vaccine is available. This form of hepatitis, which becomes chronic and debilitating, is a major job hazard for healthcare professionals, and associated personnel such as policemen, emergency medical technicians (EMTs), firefighters, and paramedics. Because these people come in contact with the blood of others and no vaccine is available to protect them from being contaminated, they must be extremely careful while dealing with patients.

- ✔ **Hepatitis D:** Usually affects people who are exposed to the blood of others on a regular basis, such as hemophiliacs (who receive blood products from other people) or intravenous drug users.

- ✔ **Hepatitis E:** The type that people acquire after having traveled to an endemic area. An "endemic" area is one where a certain disease is prevalent. For example, an endemic area for malaria is Africa. For hepatitis E, endemic areas include India, Asia, Africa, and Central America.

When one of the hepatitis viruses infects a person, the person develops flu-like symptoms in less than a week. Symptoms include fatigue, weight loss, depression, headache, aches in the muscles and joints, nausea and/or vomiting, and a generally sick feeling (given the all-encompassing name of *generalized malaise*). Nervous system problems, such as an altered sense of taste or smell and *photophobia* (when light hurts the eyes), also can occur. At this point, called the *prodromal stage,* the affected person can easily transmit the virus to others; it's extremely contagious. The prodromal stage lasts about two weeks.

Then, the real illness begins. The stage of actual illness is the *clinical stage.* During the clinical stage, the person develops abdominal pain and tenderness and indigestion as the liver enlarges from inflammation. The damaged liver can't filter *bilirubin* (see Chapter 12) from the blood, so the urine has a dark color, and the feces have the color of clay. In addition, with excess bilirubin in the blood, the skin and whites of the eyeballs take on a yellowish hue. Recovery — appropriately called the *recovery stage* — can take anywhere from two weeks to several months. The body fights the hepatitis virus like it does any other virus (see Chapter 13); medication usually isn't given. Rest and proper diet are essential, though.

The bright spot in all this is that the liver is a regenerative organ. The liver is capable of repairing and replacing its damaged cells and tissues. Damage from hepatitis usually isn't permanent. Complications, such as chronic hepatitis or liver cancer, can develop in some cases, though.

## Irritable bowel syndrome

*Irritable bowel syndrome* is one of those responses. Although the lining of the large intestine does not show inflammation — *making irritable bowel syndrome a noninflammatory bowel disease* — the tissues do become irritated, resulting in a change in peristalsis, which speeds up or slows down. Diarrhea or constipation or both can occur. Stress reduction and a high-fiber diet usually are in the treatment plan.

## Painful pancreatitis

Just as appendicitis is inflammation of the appendix, *pancreatitis* is an inflammation of the pancreas. If the inflammation occurs suddenly, the condition is *acute pancreatitis*. If the pancreas is inflamed constantly, and the function of the organ is affected, the condition is *chronic pancreatitis*. Pain associated with mild pancreatitis is centered around the navel (belly button) and does not lessen with vomiting. Usually, that type of pain is the only symptom. But with severe pancreatitis, the pain is an unrelenting, piercing pain in the middle of the abdomen. If the pancreatitis is severe and sudden, hemorrhage can occur, destroying the pancreas and resulting in acidosis (too much acid or too little bicarbonate in the blood and tissues), shock, or coma.

Acute pancreatitis can result in fluid accumulation and swelling of the tissues in the organ — a condition called *edematous pancreatitis* — or it can result in the death of cells and tissues in the pancreas — called *necrotizing pancreatitis*. The cause of these types of acute pancreatitis is the same: enzymes that are activated prematurely. But, the factors that lead to premature activation of enzymes vary from diseases affecting the bile ducts to alcoholism.

Alcoholism often leads to pancreatitis because alcohol increases the amount of juice that the pancreas secretes, changes the pancreatic cells, and hastens obstruction of ducts by making proteins clump together. When the ducts are blocked, the pancreatic juice flows back into the pancreas. Just as the pancreatic juice contains enzymes that digest food, when the pancreatic juice backs up into the pancreas, the enzymes can digest the pancreas. The tissues become damaged or die, and thus the normal functions of the pancreas are reduced.

The pancreas produces insulin, the hormone that shuttles glucose out of the bloodstream and into cells where it's converted into fuel the body can use. When the pancreas produces too little insulin, the disease *diabetes mellitus* results. With diabetes, the blood glucose level is high, yet the cells are starving for fuel (see Chapter 8). Often, after a bout of pancreatitis, the damaged pancreas does not produce enough insulin, and diabetes results.

## Ulcerative colitis

*Ulcerative colitis* (*colitis* being the inflammation of the colon) is also an inflammatory bowel disease, but it's limited to the large intestine. This condition is fairly common. Ulcers form in the mucosal and submucosal layers of the large intestine, starting in the rectum and moving upward throughout the colon. As the ulcers spread, the lining of the colon swells and begins oozing mucus and pus. Abscesses can form in the lining, and the tissue surrounding them become irritated, damaged, and die. The ulcers form periodically — they form and spread during an attack, and between attacks are periods of remission.

Because of all this damage, the feces are filled with blood and mucus. If the blood loss is severe enough, anemia can develop. Also, as the disease progresses, the lining of the colon thickens and develops scar tissue, so absorption of nutrients is reduced. Electrolyte imbalances also can occur.

The cause of ulcerative colitis hasn't been definitively decided. A problem with cells of the immune system is thought to cause the condition. The lymphocytes called T cells (the ones that are so low in people with AIDS — see Chapter 13) appear to negatively affect the cells in the lining of the colon. Another theory is that an infection may start the process. And although stress isn't responsible for causing ulcerative colitis, it certainly can lead to an attack.

Treatment during an attack includes total parenteral nutrition (TPN), which is a nutritional solution given intravenously to allow the digestive system time to heal. This means that the person with ulcerative colitis can have nothing by mouth — no food, no drinks. Medications include steroids to control inflammation, anti-diarrhea drugs, and iron supplements if anemia develops. Surgery is necessary if the TPN and medications don't work, or if the colon expands to the point where it could burst (a life-threatening condition).

# Upsetting ulcers

The wearing away of tissue can cause a sore or lesion called an *ulcer.* The digestive system is lined with (and its organs surrounded with) a mucous membrane called the *peritoneum.* The peritoneum secretes mucus not only to mix with food and create chyme, but also to protect itself from the strong hydrochloric acid in the gastric juice, which could eat the membrane away. However, sometimes the acid prevails, and the mucosal lining of the digestive tract does develop ulcers. The parasympathetic nervous system is involved in sending a signal to cause gastric juice to be secreted. So, for a long time, physicians believed that a person who was under stress secreted more gastric juice than was needed, leading to the lining of the stomach being eaten away by the excess acid.

However, fairly recently, a bacterium was discovered to have an effect on the lining of the stomach. *Helicobacter pylori,* more often than not, is the culprit in stomach ulcers. As the bacterium infects the body, it burrows into the mucosal lining of the stomach, creating an ulcer. Rather than antacids, medication, or surgery, physicians now prescribe an antibiotic to most people with ulcers.

# Chapter 12

# Cleaning Up: The Urinary System

• • • • • • • • • • • • • • • • • • • • • • • • • • • • • • • • • • • • • • • • • • • •

## In This Chapter

▶ Appreciating the difference between defecation and urination

▶ Seeing what structures make up the urinary system

▶ Figuring out how urine is made

▶ Understanding some common urinary problems

• • • • • • • • • • • • • • • • • • • • • • • • • • • • • • • • • • • • • • • • • • • •

*B*efore you start reading this chapter, I suggest that you go to the bathroom and empty your bladder. While reading about how urine is produced in and released from your body, you'll most likely feel the urge to go anyway. So save yourself some time and go pee now.

Feel better? Okay. Now you can find out about what's occurring inside your body to produce more urine while it's happening. Actually, this chapter shows what's occurring constantly in your body. You never stop making urine; you just need to stop what you're doing a few times a day to release it. Now, if you'll excuse me, I need to go to the bathroom before I keep writing.

## Doing the Dirty Work

*Urine* is the body's primary waste product. *Feces* are waste products of the digestive tract, but urine removes discarded waste products from every cell and every system of your body. So *defecation* — the release of feces — is the final step of digestion, but *urination* — the release of urine — is the final step of all metabolism.

 Together, urination and defecation are often called *excretion,* although some purists argue that only material that was used metabolically in cells is technically excreted. If that's the case, then feces is errantly called *excrement* because the matter that makes up feces was only ever in the digestive system, and never used metabolically in all the body's cells. (However, the nutrients acquired from food broken down by the digestive system — see Chapter 11 — are absorbed through the lining of the small intestine, put into the circulatory system, and then used metabolically in all the body's cells. The waste products of metabolism — the conversion of fuel to energy used in the body — go back into the bloodstream and are filtered out and then the urinary system removes them.)

This section describes the anatomic structures that make up your urinary system (see Figure 12-1), as well as the steps those organs take to produce urine. Of all the systems in the body, your urinary system has to be the most simple. There really are only four structures:

✔ Kidneys

✔ Ureters

✔ Bladder

✔ Urethra

Only one of them (the kidneys) actually produces anything. The others serve as tubes for transport or as a sac for holding what the kidney produces.

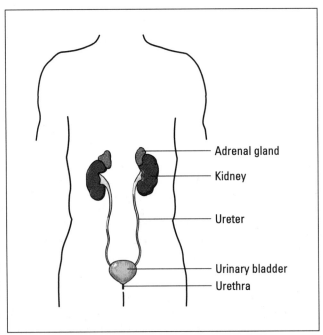

**Figure 12-1:**
The urinary
system.

Adrenal gland

Kidney

Ureter

Urinary bladder

Urethra

*From LifeART®, Super Anatomy 1, © 2002, Lippincott Williams & Wilkins*

# *Taking out the trash: The kidneys*

You may think of the bladder first when you think of organs involved in urina-tion, but the kidneys are the more important structures. Your kidneys produce urine. *Kidneys* is plural because you have two of them — one kidney on each side — located just below the ribs in your lower back (right where boxers throw "kidney punches"). Your back muscles on either side of your vertebral column help to protect your kidneys, as do your lowest ribs.

A protective membrane called the *peritoneum* surrounds the area containing most organs. Peritoneum is kind of akin to the sheathing material in which new homes get wrapped; that plastic material surrounds the area containing the framed-out rooms (like body cavities between the bones), as well as plumbing and electrical systems (like organs of your digestive, urinary, and reproductive systems). Connective tissue and adipose (fat) tissue attach your kidneys to your posterior abdominal wall — outside of the peritoneum, creating a *floating kidney.* (Adipose tissue is akin to the insulation on houses.) A hard blow to the back can dislodge a kidney fairly easily.

Most of the waste products of metabolism exit your body via your kidneys, the main excretory organs, and urinary system. Although your skin excretes sweat that contains water and salts, and your lungs excrete the waste prod-uct of respiration — carbon dioxide, and your liver removes dead red blood cells and excretes bile pigments, your kidneys filter your blood to remove the cellular wastes, as well as some water, salts, and bile pigments.

The kidneys do a good job of filtering the blood because of their structure (see Figure 12-2). Kidneys are shaped like, well, yes, kidney beans. Like the heart, each kidney is about the size of a small clenched fist. The kidneys have a reddish-brown color, like the deep reddish-brown of — believe it or not — a kidney bean.

If you happen to have a can of kidney beans in your pantry or cabinets, take out a little bean in the spirit of discovery. You can use your little kidney bean to figure out the real kidneys inside you. First, the outer covering of a kidney — the part with the reddish-brown color — comes off easily. Think of that cover-ing as your kidney's *capsule.* When you slough off the bean's covering, you see a lighter-colored meaty part inside. That part is akin to your kidney's *cortex.*

If you sliced a kidney lengthwise and opened it up (like you'd butterfly a pork chop), you'd see triangular sections that look like seashells. Blood vessels sep-arate each seashell-looking section that form the *renal medulla.* The innermost section is the *renal pelvis,* and outside of the kidney, it becomes the ureter. Both the *renal artery* and the *renal vein* enter the kidney at the point where the renal pelvis becomes the ureter. The small depression on the inner curve of the kidney where the ureter comes out and the blood vessels go in is the *hilum.*

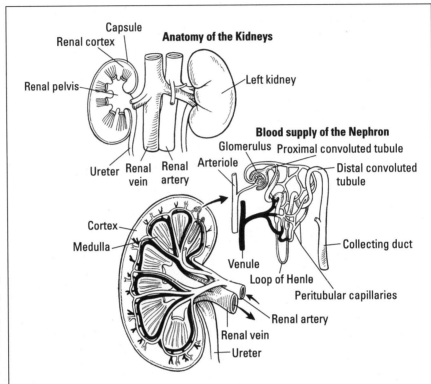

**Figure 12-2:**
Structure of
a kidney and
a magnified
look at a
nephron.

The renal artery brings blood that needs to be filtered into the kidney; the renal vein takes filtered blood out of the kidney. When blood enters the kidney through the renal artery, it flows into the renal medulla and enters a nephron (refer to Figure 12-2). A *nephron,* which can also be called a renal tubule, is a microscopic unit that filters blood and creates urine. More than 1 million nephrons occupy each kidney.

When blood comes through your renal artery, it moves into smaller branches of the artery — arterioles — which head straight into the *glomerular capsule* (also known as *Bowman's capsule*) of a nephron. Inside the glomerular capsule, the blood vessels become even smaller, forming a *glomerulus,* (clump of capillaries). The glomerulus is where your blood is filtered. As blood moves through the highly permeable capillaries in the glomerulus, the plasma (see Chapter 9) undergoes *pressure filtration*. Fluid and waste molecules such as urea and glucose are removed from the blood when the pressure that forces blood through the blood vessels is strong enough to push out water and the waste dissolved in it. (The blood pressure in the capillaries of a glomerular capsule is higher than in other capillaries in the body.) The capillary membrane basically acts like filter paper (think coffee filter), straining the fluid and removing the small waste materials. Large proteins, nutrients, and blood cells remain in the plasma.

Nephrons filter 7.5 liters of plasma every hour! That's equal to almost four big soda bottles! Men filter a little more, women a little less. But, still, that's plenty of plasma.

Substances dissolved in a solution are *solutes.* The solutes and the water in which the solutes are dissolved are carried through the tubules in the nephron. Fluids move through the *proximal convoluted tubule,* the *loop of Henle,* and the *distal convoluted tubule,* which then becomes the *collecting duct* for the nephron. Just as the intestines in the digestive system have microvilli that absorb and pass nutrients into the bloodstream, the tubules in a nephron have microvilli that reabsorb water, salts, and other needed materials back into the bloodstream via a *venule* (small branch of a vein). Active transport (see Chapter 3) passes amino acids and glucose back into the bloodstream. The active transport that occurs in the distal convoluted tubule is referred to as *tubular secretion.*

The process of reabsorption depends on what molecule is being reabsorbed, so it's called *selective reabsorption.* Sodium requires active transport to get from the fluid to the blood; sodium is reabsorbed from the proximal convoluted tubule into the *peritubular capillary* (refer to Figure 12-2). When sodium is transported out of the fluid, chloride follows. Chloride is highly attracted to sodium; they just love to be together as sodium chloride (salt).

Do you know how salt makes you retain water? Where the salts are transported actively, the water follows passively. Sodium "pulls" water out of your plasma, and the water keeps the concentration of salt balanced between the proximal convoluted tubule and the peritubular capillary. Then approximately 65 percent of the water and salts are reabsorbed from the proximal convoluted tubule and returned to your blood.

Moving salts and water in and out of tubules in the nephrons burns up about 5 percent of all your calories. If only exercise were as easy!

## Removing urea from the body's cells

*Urea* — the body's main waste product — is formed when amino acids are converted into useable energy. The amino groups from the amino acids combine with oxygen and carbon to form urea. But urea becomes toxic to cells, so your cells have to get rid of it. Urea is a metabolic waste that diffuses into the bloodstream during capillary exchange (see Chapter 9), which occurs all over your body. Then when the blood gets to your kidneys, the following gets added to your urea:

- ✔ **Salts and water:** Remaining in the tubules of the nephron

- ✔ **Nitrogenous wastes:** Metabolic waste products that contain nitrogen are gathered from all the cells around the body

- ✔ **Ions:** Atoms with positive or negative charges; examples of ions found in the body include sodium ($Na^+$), calcium ($Ca^{2+}$), potassium ($K^+$), and chloride ($Cl^-$).

## Forming urine in the kidneys

*Excretion* also can mean the movement of material from the inside of a cell to the outside of a cell, not only the release of material from the inside of the body to the outside of the body.

To understand more completely how salts, water, ions, and nitrogenous wastes combine with urea in the kidney to form urine, read through the following steps.

1. When the fluid gets to the loop of Henle, water and salt diffuse out of the ascending limb into the medulla of the kidney.

2. Depending on the concentration of sodium in the medulla, water moves into the collecting duct. (The higher the concentration of sodium, the more water moves into the collecting duct.) At the lower portion of the collecting duct, urea is excreted into the medulla of the kidney.

3. With urea diffusing out of the collecting ducts of every nephron, the medulla becomes pretty concentrated with nitrogenous compounds.

4. This high concentration attracts water; water then moves from the collecting ducts into the medulla of the kidney, where capillaries absorb it and return it to the blood.

5. The fluid and solutes that remain in the tubules of the nephron become *urine.*

Table 12-1 lists the "ingredients" of the nitrogenous wastes in urine and from where each component is gathered.

| Table 12-1 | Nitrogenous Components of Urine and Their Sources |
| --- | --- |
| *Nitrogenous Component* | *Source* |
| Urea | Amino acid metabolism |
| Creatinine | Muscle metabolism (as in digestion of meats) |
| Ammonia | Resulting from bacteria that break down proteins |
| Uric acid | Broken down nucleotides |

Urine is yellow because it contains *urobilinogen,* which is a compound formed from bacteria breaking down bilirubin in the colon. Bilirubin is produced when red blood cells are broken down in the liver (see Chapters 9 and 11). The ammonialike smell of urine comes from the fact that urine contains ammonia in addition to other substances derived from ammonia (like urea).

If you're exercising and sweating heavily, your body's fluid level is decreased. Therefore, your body holds onto the water that it has, and less water goes into your urine. Your urine becomes more concentrated and appears darker. If your body has more water than it needs, the excess water goes into the urine, making it more dilute. But as long as the urine is effectively removing all the toxins and wastes, it's doing its job.

Urine also contains *hippuric acid,* which is produced when fruits and vegetables are digested, and *ketone bodies,* which are produced when fats are digested.

People with diabetes often have excess ketone bodies in their urine because fats are broken down rather than glucose. The pancreas secretes too little insulin, so they have trouble getting glucose into their cells. In essence, the body "thinks" it's starving. And people who are literally starving also have ketone bodies in their urine. If the ketone bodies aren't excreted properly, and they build up in tissues, the body's pH level becomes more acidic, thus causing many problems with homeostasis (see Chapter 8). In the condition *ketoacidosis,* which plagues people who are starving or who have diabetes, the overabundance of ketone bodies in the tissues and fluids give their breath a fruity odor or an odor of nail polish remover (acetone).

Your body goes through the following steps to make urine:

1. **Pressure filtration** through the glomerular capsule removes substances like glucose, amino acids, salts, and nitrogenous wastes (urea, uric acid, creatinine), along with water from the blood passing through the kidney's cortex.

   During *pressure filtration,* blood moves through the highly permeable capillaries in the glomerulus and waste is filtered out through the capillary membrane. Fluid and waste are removed from the blood because the pressure forces out water and the waste dissolved in it.

2. **Selective reabsorption** takes some of the molecules (like glucose, amino acids, salts, or water) from the proximal convoluted tubule of the nephron and puts them back into the blood through diffusion or active transport.

   These substances may be needed elsewhere in the body.

3. **Tubular secretion** actively moves waste materials from the blood to the distal convoluted tubule of the nephron. Your blood is cleaned out again, removing toxins such as metabolites of ingested drugs (did you ever notice how your urine smells like penicillin when you're taking antibiotics?) and nitrogenous wastes. At this point, you have concentrated urine.

4. **Reabsorption of water** occurs in the nephron's collecting duct.

   As salts are reabsorbed to maintain balance between the kidney and the blood, water follows the salt back into the nephron. If your body needs water, such as when you're dehydrated, a more concentrated urine will emerge from the collecting duct to pass through the ureters and to your bladder, leaving more water in your body so you can re-hydrate. If your body contains excess water, the concentrated urine becomes diluted and passes through the tubes to the bladder, taking that extra water with it.

5. **Excretion** finally occurs when the concentrated urine and whatever amount of water can be released from your body is removed from your kidney to your bladder, where it's held prior to urination.

## Traveling down your ureters

When your body creates urine, and the water has been added to the "urine concentrate," it moves from the ascending loop of Henle to a collecting duct. Little by little, urine drips down the collecting ducts of the kidney's nephrons to the renal pelvis. The funnel-shaped renal pelvis funnels the collected urine into a ureter.

## Hemodialysis: A lifesaver

Because their nephrons are damaged, people with kidney disease can't filter out as many toxins or toxic wastes as necessary to stay healthy. People with kidney disease tend to produce watery urine with little waste. If nitrogenous wastes build up in the body, homeostasis is affected. With the balance of the body in jeopardy, more and more cells can become damaged or die. If enough cells die, the entire body dies, too.

To prolong the life of many with kidney disease, the blood can be taken from the body, filtered by a machine, and put back into the body. During this process called *hemodialysis,* blood is removed from an artery (usually the radial artery

in the arm), and the blood travels through a long, thin, coiled tube that's submersed in a warm water bath. The warm water helps the blood to maintain normal body temperature, and the water also serves as the medium into which the toxins diffuse. Whatever substances the patient has in excess, the water bath is missing. For example, if a patient has too much urea or potassium in their system, urea or potassium are absent in the water bath. Therefore, the blood has the higher concentration of those substances, and they diffuse out of the blood and into the water bath. This lowers the concentration of those potentially toxic substances in the blood. Then the filtered blood is put back into a vein.

Your ureters are tubes that transport the urine created in each kidney to the bladder. Each kidney has a ureter, and each ureter runs from the renal pelvis directly into the bladder. Your ureters have a structure that's similar to your intestines. Each ureter is made up of three layers: an outer covering, a muscular layer, and a mucous layer lining the tube's inside. The muscular layer contracts in waves of peristalsis, which moves the urine from the kidney to the bladder.

## *Storing urine in your bladder*

The *bladder,* the holding tank, is a hollow sac into which urine is deposited from the kidneys through the ureters. It lies in the pelvic cavity, just behind the pubic bones and centered in front of the rectum. (In females, it's in front of the uterus and vagina.) Like other organs in the urinary and digestive systems, the bladder is made up of an outer protective membrane, several layers of muscles arranged in different directions, and an inner mucosal layer. The muscle layers allow the bladder to expand and contract depending on how much urine is inside it.

The maximum amount of urine that the funnel-shaped bladder usually can hold is 600 milliliters. When urine presses against "pressure receptors" in your bladder, the pressure receptors "activate." Nerves attached to the pressure receptors send impulses to your brain as your bladder fills. The first "message" is sent when your bladder contains about 200 milliliters. When 400 milliliters of urine are in your bladder, you feel like you're going to burst. Of course, you won't, but it gets harder to control the sphincter holding the bladder shut the fuller your bladder becomes. When you're in a position to empty your bladder, the brain sends an impulse through nerves of the autonomic nervous system to relax the sphincter and contract the *detrusor muscle* of your bladder so that *micturition* — the act of urinating — can occur. Urine flows out of the bladder and down through the urethra.

### Expelling urine out your urethra

In general, the *urethra* is a tube that carries urine from the bladder to an opening (orifice) of the body during the process of micturition. In females, the urethra is about 3.8 centimeters in length, and it ends at the urethral orifice in the anterior wall of the vagina between the clitoris and the vaginal orifice. In males, the urethra is about 20 centimeters long and runs down through the prostate gland and the penis, which has an opening on the tip called the *urethral meatus*.

## Maintaining Homeostasis

Your kidneys have some important regulatory functions, without which your health is jeopardized. Yes, the kidneys remove wastes from the blood and create urine. That's one major job they have. But they also are key organs of homeostasis. Homeostasis describes the ongoing adjustments made to keep all your body's systems within normal limits. It's a process of checks and balances that helps you maintain your health. The kidneys play a big role in maintaining the proper balance between the salt and water content of your blood, and they're involved in maintaining the proper pH (acid-base) level of the blood, as well. The pH scale demonstrates the concentration of hydrogen ions in a solution using a range of zero (most acidic, most hydrogen ions) to 14 (most basic, least hydrogen ions). (For more on pH, see the "Regulating your pH" section coming up later in this chapter.)

### Balancing acts

Water is lost from your body when your urine is diluted, and your body conserves water when your urine is concentrated. Your kidneys regulate whether water your body releases or conserves water.

If you're a man, more than half of your body weight (about 60 percent) comes from the water contained in the fluid in your cells (called intracellular fluid) and all your extracellular fluid (the fluid in your tissues that surround cells, the plasma in your blood, your lymph fluid, and the fluids in your digestive tract and mucus membranes). If you're a woman, only about half of your body weight comes from water because you (and I) have more adipose tissue (fat) than men. And adipose tissue contains fewer cells (and therefore less water) than does muscle tissue. If your body loses more water than it takes in, you become dehydrated. Every day, the cells in your body produce water as they metabolize, you consume water, and some foods contain water. But also every day, you sweat, urinate, exhale breath that contains water droplets, and form feces. All these processes use water. Your body must keep a check on how much water is in the system and how much is leaving the system. If more is leaving the system than is in the system, you become thirsty. Thirst is a physiologic sign that gets you to take in more water. Your hormones control thirst.

## Monitoring your blood pressure

Your kidneys use the processes of tubular secretion and tubular reabsorption to remove and replace salts and water from your blood. But the kidneys, along with your liver and adrenal glands, which sit on top of the kidneys, also help to maintain the proper balance of sodium and potassium in your blood, which affects the blood volume, which, in turn, affects your blood pressure.

As blood is filtered through the liver (see Chapter 11), blood pressure is detected. If the blood pressure is too low in the liver, chances are that the blood won't be filtered properly through the glomerulus of the nephron when it arrives in your kidney. (Remember, pressure filtration filters the blood in the kidney initially.) So your liver can increase the blood pressure by increasing blood volume (see Chapter 9).

To increase blood volume:

1. The liver produces a protein called *angiotensinogen.*

2. In the kidneys, the nephrons check the blood volume, too. If it's too low, the nephrons secrete *renin,* which converts angiotensinogen to *angiotensinogen I.*

3. As the blood continues to circulate, the pressure is checked again in the capillaries of the lungs (see Chapter 10). If the pressure is too low, the lungs secrete *angiotensin-converting enzyme* (ACE), which converts angiotensin I to angiotensin II.

4. The presence of angiotensin II in the blood causes the adrenal glands to secrete the hormone aldosterone, which causes sodium to be reabsorbed from the nephrons of the kidneys back into the bloodstream.

5. Where salt goes, water follows. As the sodium ions are actively transported into the blood, water is reabsorbed as well.

6. The movement of water into the blood causes an increase in blood volume, which increases blood pressure. Then the blood has enough pressure to pass through the glomerulus and be properly filtered to remove toxins.

Errors that occur in this renin-angiotensin-aldosterone system can cause high blood pressure. Some medications for *hypertension* (high blood pressure) are ACE inhibitors; by keeping the lungs from secreting ACE, angiotensin I can't be converted to angiotensin II, so sodium and water are kept from being put back into the bloodstream unnecessarily. Just as there are negative effects from the kidneys not removing enough toxins if blood pressure is too low, high blood pressure has negative effects, too. Balancing the two important aspects of homeostasis is of utmost importance.

The following two hormones can also affect blood pressure:

✔ **Antidiuretic hormone (ADH)** is secreted by the pituitary gland (see Chapter 8) when sodium has been reabsorbed by the kidney's nephrons but not enough water follows. ADH causes more water to be reabsorbed from the urine being produced, which decreases the amount of urine. So when you don't drink enough water ("enough" being 64 ounces [half-gallon] sipped throughout the day) — or you drink enough but have more losses (through sweating, diarrhea, or vomiting) — the tiny gland in your brain releases ADH, which keeps your blood volume (and therefore blood pressure) at normal levels.

✔ **Atrial natriuretic hormone (ANH)** has the opposite effect as ADH. When blood passes through the heart, the stretch receptors inside the heart detect the blood volume. If the blood volume is too great, some of the water needs to be taken out of the blood to reduce the volume. Cells in the heart release ANH, and it prevents the kidney from secreting renin. Without renin, angiotensin cannot be converted to angiotensin I. Therefore, ACE has no angiotensin I to act on, and angiotensin II can't be produced. The actions of ANH keep sodium and water from being reabsorbed from the kidney back into the bloodstream. Instead, sodium and water move from the bloodstream into the kidney, which decreases blood volume and subsequently blood pressure.

## Drying up hangovers

In addition to affecting your brain, alcohol is a natural diuretic because alcohol prevents the pituitary gland from secreting antidiuretic hormone (ADH). Your kidneys get to the point where sodium is reabsorbed but not enough water follows. Without ADH to cause water to be reabsorbed from the urine into the bloodstream, you urinate more but decrease your blood volume. You become dehydrated, and the resulting dehydration causes the dizzy, headache-y feeling of a hangover. To prevent a hangover, the advice is simple: Avoid alcohol. But, if you're going to drink alcohol, at least mix it with water or drink a glass of water for each glass of wine, beer, or liquor you down. If you don't follow that advice, drink water to alleviate the dehydration that accompanies a hangover.

## *Regulating your pH*

Acids contain more hydrogen than hydroxide, and bases contain more hydroxide than hydrogen. Acids have a pH (a measure of hydrogen ion concentration) from 0 to 7.0; bases have a pH of 7.0 to 14. (For more on ions, see Chapter 5.) The blood of a human needs to be right around 7.4; the normal range isn't broad. If the pH drops below 7.0, to an acidic level, the condition is *acidosis.* If the pH goes above 7.8, to a more basic level, the condition is *alkalosis.* Both situations can be life threatening. Buffers, which absorb excess hydrogen or hydroxide ions, must be added to the blood to correct both of these conditions.

Your body produces acids during metabolism, and some acids (like citric acid) are ingested. Even the digestion of fats produces acids — fatty acids. All acids contain hydrogen, so they're capable of increasing the hydrogen content — and acid level — of your blood. To deal with these acids, your body has buffer systems, and the kidneys are part of them.

The most commonly used buffer system is one that involves carbonic acid (a weak acid) and sodium bicarbonate (the sodium salt of carbonic acid). This system works in the plasma. When an acid with a lower pH than carbonic acid (that means it's stronger than carbonic acid) enters your bloodstream, from metabolism or ingestion, it's buffered by the sodium bicarbonate. I'll call the acid that's stronger and added to the system acid X. The sodium bicarbonate reacting with acid X removes some of the acid X's hydrogen, forming carbonic acid (the weaker acid) and the sodium salt of acid X.

If you drink a glass of orange juice, you're adding citric acid to your system. When citric acid gets into your bloodstream, sodium bicarbonate reacts with it, removing some of the hydrogen molecules from the citric acid, which creates sodium citrate and carbonic acid. Having the weaker carbonic acid in your bloodstream keeps the blood's pH level lower than it would be if your bloodstream contained an abundance of citric acid.

In the kidney, hydrogen ions are often exchanged for sodium ions. The sodium is absorbed by the nephrons and returned to the bloodstream, and the hydrogen is put into the urine and excreted from the body. Carbonic acid breaks down into carbon dioxide and water, both of which can go back into the bloodstream. The carbon dioxide diffuses out of the bloodstream in the lungs and is exhaled. Water, if not needed in the body, is removed in the urine.

Your kidneys also can detect if the pH of your body fluids is too low (meaning acidic). If your body fluids are too acidic, the amino acid glutamine is broken down. During the metabolism of glutamine, ammonia is produced. When ammonia is transported into the filtrate that becomes the concentrated urine, sodium ions move back into the bloodstream, where they can again form sodium bicarbonate and continue the buffer system.

# Urinary System Diseases and Disorders

Although the urinary system has few parts, and the way the parts work together isn't really all that complex, things can and do go wrong. Because of the importance of the kidneys in homeostasis, kidney problems can cause problems in other systems. The urinary system also can be attacked by bacteria, or nearby anatomic structures can put the squeeze on the urinary structures.

## Urinary tract infection

Women tend to have far more *urinary tract infections* (UTIs) than men do for the simple reason that the urethra in females is much shorter than in males. With a shorter path to travel, bacteria can go up into the bladder of women much easier than they can get to the bladder of a man.

UTIs come on suddenly and are hard to miss. The burning pain when you have to urinate is a huge sign. A UTI is a general name for an infection that can affect any structure in the urinary system: an infection in the bladder is *cystitis;* an infection in the kidney is *pyelonephritis;* infection in the urethra is *urethritis.*

Bacteria are usually the culprits in UTIs. The bacteria are often spread from the rectum to the urethra. Sometimes, sexually transmitted diseases, such as gonorrhea or chlamydia, may lead to an infection in the urinary system.

Because UTIs are so noticeable and uncomfortable, people often seek treatment immediately, which keeps the infection from spreading higher up into the urinary system. But, if the infection gets into the bladder and then spreads up into the kidneys, kidney damage can result. Because of the close association of the kidneys with the bloodstream, *septicemia* — a spread of infectious organisms through the bloodstream — can be a complication. Septicemia can lead to *septic shock,* which can be life-threatening. So if you experience painful urination, seek medical treatment.

When you consult a physician about painful urination, the physician asks for a urine sample that's tested to see what organisms are harboring within (tests performed on urine are given one term: *urinalysis*). If bacteria are present, an antibiotic is prescribed. In addition, nonprescription medications are now available to stop the pain of a UTI. But, you will still need to get a prescription for an antibiotic if bacteria are making you wince and burn.

## Calculating kidney stones

The proper name for kidney stones is *renal calculi.* Calculus (singular) means "pebble," but I think it also means "hard." The material that's scraped off your teeth by your dental hygienist is called calculus, and just look at the math subject by the same name — hard fits!

Kidney stones may result from chronically concentrated urine. People who don't drink enough water are constantly mildly dehydrated, making their urine more concentrated. The higher concentration of solutes in the urine make it easier for material, such as uric acid, to precipitate (clump). Uric acid crystals can develop into kidney stones, or they may travel to your joints, causing *gout.*

One of the most common components of kidney stones is *oxalate,* which is found in green, leafy vegetables, such as spinach, as well as coffee and chocolate. When combined with calcium, oxalate does not dissolve well, and it crystallizes into a kidney stone. Metabolic disorders or chronic UTIs can also cause kidney stones.

A kidney stone that blocks a ureter causes sudden, sharp pain that may cause nausea or vomiting. The stone may cause the lining of the ureter to bleed, producing the condition of *hematuria* (blood in the urine). If the kidney stones are smaller than five millimeters in diameter, physicians tell you, "They will pass." You must drink plenty of water to help flush the stones out. However, if they're larger, there are noninvasive techniques (that is, no cutting open the abdomen for surgery) that can break up the stones using shock waves (shock wave lithotripsy) or ultrasound waves (ultrasonic lithotripsy).

## Prostate problems

In men, the prostate gland surrounds the urethra; the urethra passes right through the prostate gland. When men ejaculate, the seminal fluid passes out of the urethra in the penis. Yes, in men, the urethra carries urine and semen out of the body, but at different times (see Chapter 14). As the seminal fluid is being produced, the prostate gland adds some secretions to the fluid on its way out of the penis.

As men age, the prostate gland enlarges, usually starting around age 50. If the prostate gland enlarges too much, it can press on the urethra and block the flow of urine. The blockage can lead to UTIs in men, and it often causes painful urination (technically called *dysuria*). As the bladder becomes stressed from the blocked urethra, urine may leak out, or, more commonly, the urge to urinate becomes more frequent.

An enlarged prostate can be detected during rectal examinations. A distended (overstretched) bladder can be felt during an abdominal examination. Treatment — usually removal of the prostate gland — is required if the symptoms are severe or if cancer is detected.

## Incontinence across the continents

Incontinence is the inability to control the release of urine. There are four different types of incontinence, but their causes are disease of the urinary tract or injury to the urinary tract.

- **Stress incontinence:** This type of incontinence is most common and is the type in which urine leaks out when stress is placed on the bladder, such as when a person runs, coughs, lifts something heavy, or laughs. This type usually affects women who have given birth several times because the muscle supporting the bladder weakens and the sphincter muscle of the urethra stretches.

✔ **Urge incontinence:** This type of incontinence is most embarrassing. It is the type in which the need to urinate ("gotta go!") comes on suddenly and the bladder empties itself, even if you are not in a socially acceptable place in which to do so. The adult "undergarments" industry is targeted to people with this type of incontinence.

✔ **Overflow incontinence:** Some people may have trouble emptying their bladders, such as those with enlarged prostate glands or other obstructions of the urethra. When something is blocking the urethra, like a kidney stone, urine backs up in the bladder. The bladder retains urine constantly because it is unable to empty totally. Instead, the bladder overflows with urine, and urine continually dribbles out of the urethra. After the blockage is removed, the urine can be drained from the bladder, and the incontinence problem usually goes away.

✔ **Total incontinence:** Some structural problems, spinal injuries, or diseases can cause a complete lack of bladder control. The sphincters of the bladder and urethra don't operate, leading to urine flowing directly out of the bladder as it's deposited into the bladder from the kidneys.

Depending on the cause of incontinence, exercises to strengthen the muscles of the pelvic floor, surgery, medications, or medical devices can help to deal with untimely urination.

# Chapter 13

# Fighting Fairly: The Immune System

. . . . . . . . . . . . . . . . . . . . . . . . . . . . . . . . . . . . . . . . . . . . . . . . . . . . . . . . .

### In This Chapter

▶ Seeing how your lymphatic system helps to boost your immunities

▶ Checking out your immune system's many different cells

▶ Understanding inflammation and your immune system's reactions

▶ Exploring the secrets of the complement system and antibodies

▶ Finding out about the diseases and disorders of the immune system

. . . . . . . . . . . . . . . . . . . . . . . . . . . . . . . . . . . . . . . . . . . . . . . . . . . . . . . . .

The human species wouldn't have survived long if the body had no immune system. The first cold virus would've made the original members of our fair species extinct right back then and there if the body's immune system didn't fight off that invading cold virus.

This chapter takes you on a tour of your immune system and shows you how your immune system protects your cells. Your immune system — your body's defense department — is active in times of health and illness. Your immune system is your body's way of protecting itself from invading microbes, such as bacteria and viruses, other foreign cells, and your own cells that have gone bad (such as cells that have become cancerous).

In addition to the immune system, your body has several ways of protecting itself.

- ✔ **Skin:** This barrier keeps out innumerable invaders. Glands in the skin secrete oils that make the barrier even more effective. (For more on the skin, see Chapter 6.)

- ✔ **Mucous membranes:** This lining in the organs of the respiratory and digestive systems traps microbes, which are minute organisms capable of causing disease in animals (including humans).

✔ **Cilia:** These tiny hairlike structures in the respiratory and digestive tracts physically move trapped dirt and microbes into the throat, where they are then swallowed and removed through digestion and excretion. (For more on cilia, see Chapter 2.)

✔ **Hydrochloric acid:** This acid in the stomach destroys most of the bacteria that may be ingested. The "good" bacteria that normally live there (the normal flora) battle the "bad" bacteria that do make their way into the intestines. (See Chapter 11 for more on hydrochloric acid.)

All these mechanical processes in the preceding list help to keep microbes out of the bloodstream and tissues, but a defense system — your immune system — combats those microbes that do make it into the inner territory.

Your actual immune system consists of a variety of cells and some molecules. Unlike other systems, the immune system doesn't have a group of organs that work together in a centralized location. Instead, your immune system uses pathways in the lymphatic and circulatory systems to transport immune system cells around your body. (See Chapter 9 for more on the circulatory system.) These immune system cells go on scavenger hunts for germs that don't belong inside your body. Then, they attack the foreigners, and disable, kill, or remove them to keep you healthy. (For more on the immune system's cells, see the "Scavenging and Attacking: Immune System Cells" section, later in this chapter.)

# *Loving Your Lymphatic System*

The lymphatic vessels — the tubes that carry lymph fluid throughout your body — have a structure similar to the veins of your circulatory system (see Chapter 9). Veins have valves, which prevent deoxygenated blood from flowing backward. Likewise, the lymphatic vessels contain valves that keep lymph from flowing backward. (See sidebar "What is lymph, anyway?," for a description of lymph.) Therefore, the lymphatic system is a one-way system. Both veins and lymphatic vessels need to be squeezed to get the blood or lymph moving through them. You squeeze them whenever you move your arms, legs, and other skeletal muscles.

Extra tissue fluid is absorbed by the lymphatic vessels and returned to the bloodstream as lymph.

Like the circulatory system, the lymphatic vessels branch out and increase in size from capillaries to larger lymphatic vessels. These larger vessels enter the thoracic duct (also called the left lymphatic duct) or the right lymphatic duct of the lymphatic system. The right lymphatic duct, which is located on the right side of your neck near your right collar bone, takes lymph that

## What is lymph, anyway?

Lymph, the fluid that flows through lymphatic vessels, basically is extra tissue fluid. Tissue fluid (also called interstitial fluid) surrounds every cell. The cells need to be in a fluid environment because nutrients and oxygen need to be dissolved in the fluid in order for them to move into the cells. When an excess of tissue fluid is created, the extra fluid is absorbed into the lymphatic vessels and is then called lymph.

The lymph courses through the lymphatic vessels, carrying with it some lymphocytes (white blood cells), occasionally a few red blood cells, and sometimes fat particles. As lymph passes through lymph nodes, the lymph gets filtered to remove microbes or impurities. Then, the purified fluid is returned to the bloodstream via the veins of the circulatory system.

drains from the right arm and the right half of the body above the diaphragm. The thoracic duct, which runs through the middle of your thorax, takes lymph that drains from everywhere else in the body. Lymph is deposited into the circulatory system via the subclavian veins, located right near the heart and major blood vessels, so you can see how germs from an infection have the potential to spread through the bloodstream or to the heart. The right lymphatic duct enters the right subclavian vein, and the thoracic duct enters the left subclavian vein.

Before lymph drains into ducts and gets into the circulatory system, it passes through lymph nodes. A lymph node is an encapsulated mass of tissue that filters the lymph just before it enters the bloodstream. Fibrous connective tissue (see Figure 13-1) both encapsulates (covers) lymph nodes and separates nodes into *nodules* (smaller parts of nodes). Cute name, isn't it? The outer part of a nodule — the cortex — contains two types of *lymphocytes* (immune system cells):

✔ *B lymphocytes:* Developed in an area of the cortex called *the germinal center* (the center of a lymph nodule), these cells produce antibodies. (See the "Antibody-mediated immunity" section, later in this chapter, for more on antibodies.)

✔ *T lymphocytes:* Developed in areas of the cortex other than the germinal centers, these cells are responsible for killing other cells that are harboring a virus. Rebellion is not allowed at any level!

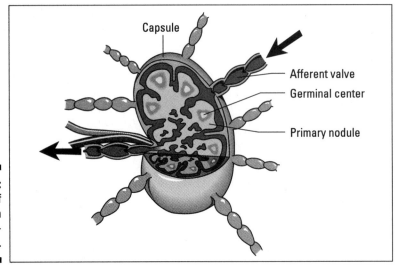

Capsule

Afferent valve

Germinal center

Primary nodule

From LifeART®, Super Anatomy 3, © 2002, Lippincott Williams & Wilkins

**Figure 13-1:**
Anatomy of a lymph node, cross-section.

Lymph nodes are found all over your body (see Figure 13-2), under your armpits, in your neck, and in your groin. The neck and mouth contain a large amount of lymph nodes because with all the *orifices* (openings) in your face — your nostrils, ears, mouth — this area is a prime location for bacteria and viruses to enter your body. The *tonsils* in your throat (pharynx) are large lymph nodules, and these pharyngeal tonsils are also *adenoids*. But you also have tonsils at the back of the oropharynx *(palatine tonsils)* and at the base of the tongue *(lingual tonsils)*.

Having lymphoid tissue in the form of tonsils in and around your throat is like having armed guards standing at the entrance to the inside of your body: They trap and remove many potentially harmful microbes. Another prime location for microbes to get the best of you is your intestines. But your intestines have a little outpost there, too: *Peyer's patches* are lymph nodules found in the wall of the intestines.

If you've had chickenpox, a cold, or the flu, a virus attacked you. A *virus* is an infectious agent that lacks the ability to metabolize or reproduce; therefore, it needs to "take over" a cell that has those abilities in order to survive and produce offspring. As an "infectious agent," a virus is not working undercover in a beige overcoat, tiny sunglasses, and a wide-brimmed hat. But, a virus technically is not a cell because it does not have a nucleus or other cellular organelles (which is also why it cannot reproduce or metabolize). Your T cells fight the virus directly; your B cells form antibodies, which remain in your system to be ready at a moment's notice the next time that same virus re-enters the scene. (For more on antibodies, see the "Antibody-mediated immunity" section, later in this chapter.)

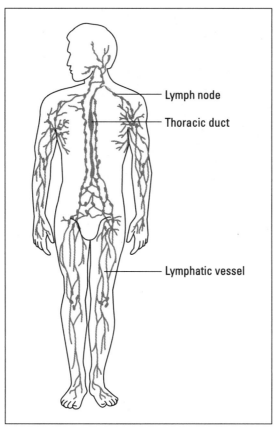

*From LifeART®, Super Anatomy 3, © 2002, Lippincott Williams & Wilkins*

**Figure 13-2:**
The
lymphatic
system.

When a virus starts attacking your cells, your immune system springs into action to protect you as follows:

1. Your lymph nodes begin producing more B and T lymphocytes.

2. The T lymphocytes in your lymph nodes start killing off the cells that contain the virus (this also makes you feel run down).

3. B lymphocytes make *antibodies* to combat each *antigen* (a foreign cell, like a virus) that enters your body. (See the "Defending Your Health" section later in this chapter for more on antibodies and antigens.) Some antibodies stay in the body for a short time, some for several years, and some for the rest of your life. Whenever a virus that you've had before enters your system, your body immediately recognizes it as a previous invader and eliminates it before it can make you sick again. Antibodies are the reason why you don't get chickenpox if you've had chickenpox, and it's also the reason why people say you never get the same cold twice.

## Falling victim to a vicious virus

If you've had chickenpox, the chickenpox virus got into your body through the water droplets from someone else's sneeze or cough. Then to survive, the virus entered your cells and used your genetic equipment to replicate and your organelles to metabolize. As more and more of your cells were taken over, you felt sicker and more tired because the chickenpox virus took your nutrients and oxygen to produce energy for itself. You probably had the usual telltale signs of chickenpox: flulike symptoms, red pustules that itched like crazy, and fever. Well, those "signs" are really reactions to help eliminate the nasty virus. For example, an increase in mucus production causes a runny nose in an attempt to flush virus particles out of the nasal cavities. A fever raises your temperature in an effort to make your body an inhospitable place for a virus to live.

## *The splendid spleen*

Your spleen lies in the upper left quadrant of your abdomino-pelvic cavity (see Chapters 1 and 5), right underneath your diaphragm. It serves as a home for cells of the immune system. Essentially, its structure is that of a lymph node — a really large lymph node. Your spleen contains connective tissue and is divided into lobules (smaller sections within a lobe) that contain red pulp and white pulp:

- **Red pulp:** Red because it contains red blood cells, red pulp also contains white blood cells, including lymphocytes and macrophages.
- **White pulp:** This pulp contains the lymphocytes and macrophages, but no red blood cells.

The lymphocytes and macrophages are immune system cells that help to purify the blood by removing microbes or other foreign material.

Just as lymph nodes filter lymph to remove impurities, the spleen filters blood. Although your spleen can be removed without killing you, people who have their spleens removed are often more susceptible to infections because they no longer have the lymphocytes and macrophages filtering microbes out of their blood.

Your spleen also stores any excess blood and then releases it whenever it's necessary to increase the amount of blood flowing through your vessels (such as when your blood pressure is low or when your body needs more oxygen).

## The thymus gland puts the "T" in T cells

The thymus gland overlays the heart and straddles the trachea, sitting just behind (posterior to) the sternum. The thymus is largest in children; it decreases in size with age. In adults, the thymus is small, but it does have a purpose — it produces *thymosin,* a hormone that stimulates the differentiation and maturation of T cells. (See Chapter 8 for more on hormones.)

When lymphocytes are "born" in the cortex of a lymph nodule, they're just plain ol' lymphocytes that eventually differentiate into B or T lymphocytes. (Refer to more on lymphocytes in the "Loving Your Lymphatic System" section, earlier in this chapter.) Before they differentiate, the young lymphocytes have some interaction with the body's cells — a kind of audition. If the young lymphocytes attack any of your body's own cells, the lymphocytes die. The lymphocytes that don't attack your body's cells, but rather show an enthusiasm for going after cells that are foreign, mature into full-fledged fighting T lymphocytes and move to the *medulla* (middle) of the lymph node. In the medulla, the fighting T lymphocytes further differentiate into one of several different cell types: helper T cells, cytotoxic (cell killing) T cells, or suppressor T cells. By default, the remaining well-behaved lymphocytes that don't attack the body's own cells or enthusiastically go after foreign cells become the B lymphocytes that form antibodies.

# Scavenging and Attacking: Immune System Cells

The cells of your immune system are the white blood cells that float through your circulatory system, so you can see how closely the immune system is connected with the circulatory system. But the immune system also has a connection to the skeletal system, your bones. Read on.

All blood cells — both red and white — are created in the red marrow of your bones. Red bone marrow is found in the ends of long bones, such as the femur in your thigh and the humerus in your arm, as well as in the ribs, skull, pelvis, spinal column, clavicles (collar bones), and sternum (breastplate).

Within the red bone marrow are specialized cells called *reticular cells,* which produce the meshlike connective tissue fibers in the bone, and *multipotent stem cells,* which have the potential to become many different types of cells (hence the name multipotent): a red blood cell, a platelet, or one of five types of white blood cells (Table 13-1). Both reticular cells and multipotent stem cells line the sinuses in veins that run through the canals in the Haversian systems of bones (see Chapter 4). As the multipotent stem cells differentiate into a specific type of blood cell, they pass into the bloodstream through the veins.

Cells are constantly growing and dividing. When a multipotent stem cell divides, it produces a *myeloid stem cell* and a *lymphoid stem cell.* As a myeloid stem cell grows and divides, it's capable of producing three types of cells:

- ✔ **Erythroblasts:** Erythroblasts mature into erythrocytes (red blood cells).

- ✔ **Megakaryoblasts:** Megakaryoblasts mature first into *megakaryocytes,* which then break apart into *thrombocytes* (platelets).

- ✔ **Myeloblasts:** Myeloblasts mature into one of four types of the white blood cells (*leukocytes*): basophils, eosinophils, neutrophils, monocytes.

    The basophils, eosinophils, and neutrophils are considered *granular leukocytes* because when these cells are stained, tiny granules in their cytoplasm are easily seen. Monocytes, although they also have granules in their cytoplasm, they are less easily seen, so they're referred to as *agranular leukocytes.*

B and T lymphocytes are also agranular leukocytes, but the original multipotent stem cell produces a lymphoid stem cell that the B and T lymphocytes develop from. The B cells leave the bone marrow as functioning lymphocytes. The T cells must go to the thymus gland first before they become fully mature and functioning immune system cells. (See the preceding section, "The thymus gland puts the "T" in T cells.")

| Table 13-1 | Cells of the Immune System | |
|---|---|---|
| *Cell Type* | *Origin* | *Characteristics* |
| **Granular leukocytes** | | |
| Basophil | Myeloblast: Originates from myeloid stem cell that comes from multipotent stem cell in red bone marrow. | Releases histamine. Makes up only about 1% of total white blood cells. |
| Eosinophil | Myeloblast: Originates from myeloid stem cell that comes from multipotent stem cell in red bone marrow. | Phagocytizes (engulfs) complexes of antigens and antibodies, then splits them apart. Makes up 1–4% of total white blood cells. |

| Cell Type | Origin | Characteristics |
|---|---|---|
| Neutrophil | Myeloblast: Originates from myeloid stem cell that comes from multipotent stem cell in red bone marrow. | Phagocytizes bacteria it encounters while traveling through bloodstream and body. Makes up the majority (40–70%) of total white blood cells. |
| **Agranular leukocytes** | | |
| Monocyte | Myeloblast: Originates from myeloid stem cell that comes from multipotent stem cell in red bone marrow. | Develops into a macrophage, which phagocytizes both bacteria and viruses. Makes up 4–8% of total white blood cells. |
| B lymphocyte | Lymphoid stem cell: Originates from multipotent stem cell in red bone marrow. | Produce antibodies. Combined, B and T lymphocytes account for 20 – 45% of total white blood cells. |
| T lymphocyte | Lymphoid stem cell: Originates from multipotent stem cell in red bone marrow; matures in thymus gland after stimulation from the hormone thymosin. | Destroy cells containing viruses. Combined, B and T lymphocytes account for 20% to 45% of total white blood cells. |

# Inflammation Is Swell

If you've ever had a splinter, you may have noticed that the site of entry of that little shard of wood became swollen, red, and tender. The swelling and redness is part of an *inflammatory reaction*. The pain draws your attention to the site so that you can take care of yourself. Then you remove the offending object and clean the wound.

Because capillaries are so pervasive in the body, when tissues are injured, capillaries are likely to be injured as well. A damaged capillary triggers the release of *histamine,* a chemical produced in all tissues of the body. (Antihistamine medications inhibit the production of this chemical that's a key in allergic reactions.) Histamine causes dilation of the capillary, which allows a larger volume of blood to pass through the capillary, making the area look red. The more blood that floods the area, the more immune system cells are brought to the site to fight any microbes that were carried in on the shard of wood or that were able to get in through the opening in the skin.

When histamine is released, the chemical *bradykinin* also is released. Bradykinin causes the nerves to send impulses to the brain that are interpreted as "Ouch!" But bradykinin makes the capillaries even more permeable, thus more tissue fluid is allowed to pass into the wound. The extra tissue fluid helps to flush out any sources of potential infection, but it also results in swelling. The increased permeability allows the immune system cells to easily pass into the wound. And any excess tissue fluid is removed as lymph and filtered in lymph nodes. Neutrophils and monocytes (refer to Table 13-1) usually are the first line of defense against any little invading buggers.

## Neutrophils to the rescue!

Like a garbage disposal, the neutrophils gobble up a bacterium (*bacterium* is singular form of "bacteria") through the process of *phagocytosis*. (*Phago-* means "eating"; *cyt-* refers to "cell.") When a neutrophil "eats" a bacterium, it surrounds the bacterium with a vacuole (see Chapter 3), which fills with enzymes that split the bacteria apart. (Enzymes are proteins that speed up a chemical reaction. For more on enzymes, see Chapters 1 and 11.) The granules in the neutrophil spark the reaction of the enzyme acting on the bacteria.

## Monocytes can really fight

The monocyte grows into a big, ugly monster cell called a *macrophage*. Macrophages scavenge for and then eat many, many microbes. When an infection is increasing, macrophages initiate the production of more white blood cells — in essence, macrophages build up the troops. This "building up of troops" explains why an increased white cell count is an indication of a growing infection within the body.

# Defending Your Health Against Invasion

In times of health, skirmishes and small battles are constantly occurring in your bloodstream and tissue — your immune system ambushes or fends off those barbarian viruses and bacteria when they first come into view. When

COOL BODY BITS

## What's all the pus about?

"Eeeww, pus!" "Yuck, that's gross!" Pus is another good thing that comes from your immune system. It may not be one of Martha Stewart's "good things," but it's a good thing for your body if you're being infected.

Pus contains neutrophils that die after phagocytizing bacteria, some dead cells and tissue, trapped bacteria, and white blood cells. All that material together forms the thick, yellow substance that people think is so gross. But pus is proof that your body is fighting an infection and that your immune system is doing its job. Be thankful for pus!

you're sick, however, the war escalates. The artillery used in these battles includes different types of proteins. One group of proteins that travel in the plasma of your blood is the *complement system*.

## The complement system

The complement system is complex, but basically works like a chain reaction. When one complement protein is activated, it then activates another one, and so on. The chain of events that occurs in the complement system begins when microbes are detected in the body. Several different complement proteins work together to make the complementary system.

- ✔ Some of the complement proteins are activated to produce additional proteins to degrade holes in the cell membrane of a microbe.
- ✔ Other complement proteins cause the release of chemicals that serve as signals for inflammation and phagocytosis to begin.
- ✔ Still other complement proteins bind to a microbe that's covered with antibodies; in essence, they "mark" the microbe for death because the immune system cells that phagocytize the bacteria are attracted to those complement proteins.

Two additional methods of defense include antibody-mediated immunity and cell-mediated immunity.

# Antibody-mediated immunity

Antibody-mediated immunity is the defense method that uses antibodies, developed from B cells, as the troops that go into battle.

An *antigen* is any cell or protein that your body thinks is foreign (your body didn't make it) and thus wants to fight off. Material that your own body produces is referred to as *self*. Anything that's not made in your body is considered *nonself*.

Antigens can be

- ✔ Proteins on a bacterium or virus membrane.
- ✔ Part of a nonself cell.
- ✔ Cancerous cells.

B lymphocytes produce *antibodies* — globular proteins also known as *immunoglobulins* — that combine with antigens, forming antigen-antibody complexes. (Refer to the "Loving Your Lymphatic System" section earlier in this chapter for more on B lymphocytes.) When an antibody combines with an antigen as an antigen-antibody complex, the antibody renders the antigen inactive and unable to attach to a cell and take it over. The antibody sends out more antibodies against that specific antigen through the blood, lymph, and mucus.

The antigen-antibody complex also calls the immune system cells into active duty. Neutrophils or monocytes, which then become hungry macrophages, usually come to gobble up the antigen. Or the complement system may be activated, thus causing the complement proteins to attract the phagocytizing cells (macrophage cells that break down and consume the antigen). To protect you from further infection, antibodies saturate your body's cells and tissues shortly after bacteria or a virus invades while your immune system is killing and removing the invaders.

Each time that your body encounters a new antigen, your B lymphocytes produce antibodies specific to that antigen. The surfaces of lymphocytes are covered with receptors, which are molecules that fit with a specific antigen. The receptor fits with the antigen like a key fits in a lock. Then, when the antigen is encountered again, the B lymphocytes are able to recognize it. The more antigens that are encountered, the more receptors are developed, making you immune (or giving you immunity) to more antigens.

Although you can produce a million or so antibodies during your lifetime, antibodies are categorized into five classes — basic structures with certain characteristics; a million or so varieties can be created because each antibody has a *variable region* in its structure that becomes specific for each encountered antigen. Remember that antibodies are also called immunoglobulins, which are abbreviated "Ig."

- ✔ **IgA:** This class of antibodies works against the toxins that bacteria produce, and they directly attack some microbes. IgA antibodies are found in secretions (like breast milk and saliva), and they protect the body cavities.

- ✔ **IgD:** Receptor sites on B lymphocytes, these antibodies are found primarily in the blood and lymph.

- ✔ **IgE:** These antibodies cause allergic reactions and are found in attack parasites such as worms. They're found in the membranes of basophils, which travel in the blood, and of mast cells, which reside in tissues.

- ✔ **IgG:** The most common antibodies found in the blood and lymph, they directly attack microbes and the toxins of bacteria, and they step up phagocytosis. IgG antibodies are the type most involved in fighting an antigen the second time it appears in the system (secondary response); therefore, these antibodies are a good measure of the strength of your immune system and your ability to resist certain diseases.

- ✔ **IgM:** These are the largest antibodies traveling in your blood and lymph. They are the first antibodies to show up after a microbe starts infecting you (primary response), and they activate the complement system.

## Cell-mediated immunity

Cell-mediated immunity is another method of defense, but instead of using antibodies (as in antibody-mediated immunity) as the battling troops, it uses T cells.

COOL BODY BITS

## Making the perfect match

Killer T cells attack a donated organ if its *major histocompatibility* (MHC) *proteins* (*histo-* means tissue; *compatibility* refers to alikeness) are recognized as foreign. For successful organ transplants, a donor's organ(s) need to be composed of tissues that match the tissues in the recipient's organ(s). The MHC proteins determine the "match." The need to match is so important because if the T cells in the recipient "decide" that the donor tissue is foreign, the T cells treat the donor organ as an antigen and launch an attack.

B lymphocytes produce antibodies. But besides B lymphocytes, the lymph nodes also mature the T lymphocytes (called T cells). T lymphocytes must pass through the thymus to mature fully. After they mature, though, look out. T cells are extremely important to your immunity. If your body's initial barriers, such as skin or mucus, get passed by a microbe that can cause an infection, the onus falls to the cells of your immune system to protect you. T cells are the army troops that do the fighting, whereas the B cells antibodies are the armed guards that hold a grudge against past offenders.

T cells become activated when a macrophage or Langerhans' cell (in the skin) that has captured a microbe "shows" the T cell the microbe's antigens. This action is *presentation,* and it helps determine what type of T cell the lymphocyte will turn into. T lymphocytes can differentiate into two distinct types of T cells.

- **Helper T cells:** These T cells help the immune system protect you. They release proteins that stimulate B cells to produce antibodies that stimulate other cells of the immune system to mount an attack.

- **Killer T cell:** When these killer T cells (also called *cytotoxic T lymphocytes*) bump into an antigen during their passage through the bloodstream or lymph, they attack and kill right on the spot: They release chemicals that bore into the foreign cell's membrane. The foreign cell's cytoplasm and contents burst out the hole(s) in the cell that the chemical created and the foreign cell dies.

  Killer T cells also kill any cell in the body that contains a virus (so the more killer Ts you have, the more resistant you are to viruses), and they go after cells that have become cancerous, because the changes in a cell that render it cancerous usually make the cell no longer recognizable as "self."

  Mutations can lead to altered cells that have the potential to become cancerous. However, if your killer T cells are on top of their game, they eliminate these potentially cancerous cells. The problem with cancer occurs when the abnormal cells begin dividing and multiplying faster than the killer T cells can. Then the killer T cells cannot eliminate all the abnormal cells, and the cancer can grow and spread.

T cells are extremely important to your immunity. If a microbe that can cause an infection infiltrates the body's initial barriers, such as skin or mucus, the responsibility falls to the cells of the immune system to protect you. T cells are the army troops that do the fighting, whereas the B cell's antibodies are the armed guards that hold a grudge against past offenders. When a T cell encounters an antigen, the T cell perks up and lets its receptor and a protein on its surface make contact with the offender. The protein on the surface of the T cell is a *major histocompatibility* (MHC) *protein.* (See the sidebar "Making the perfect match" for more on MHC.) Your genes determine the types of MHC proteins that exist on your T cells.

# Granting Immunity

Medical research and technology have given the human body's immune system an edge. With man-made products (sometimes with the help of other animals) called *vaccines,* immune systems can be given a "heads up" and a head start on forming antibodies against certain nasty antigens. Prior to the development of these products, however, people had to boost their immunity the old-fashioned way — they had to get sick. They died from it or survived it. If our forebears survived the illness, then they developed a natural immunity to the disease. In the past, many people died from illnesses that you're protected from today.

## Vaccines

Vaccines are not shots. *Vaccines* are mixtures containing material that are considered antigens: killed microbes, material from a live microbe that cannot cause infection, or certain proteins that the body recognizes as foreign. Vaccines are administered either orally (by mouth), nasally (by spray through the nose) or by inoculation (needle through the skin). *Inoculations* usually are intramuscular (into the muscle) or subcutaneous (in the subcutaneous layer of the skin).

Scientists have to make viruses unable to cause infection but keep their proteins so that they're recognized as an antigen in order to initiate a reaction by the immune system. This is called creating an *attenuated virus.* The virus is grown (cultured) under adverse conditions. The virus "looks" the same, but doesn't act the same — it has lost its virulence.

Killed viruses are also used for some vaccines, although they're usually not as effective as attenuated viruses. Treating the virus with high heat or chemicals, such as formaldehyde or phenol, typically kills the virus. The dead virus cells are injected into the body, and the immune system creates antibodies against the foreign cells.

Because vaccinations require an action by your immune system — the active development of antibodies — they're said to bestow *active immunity.*

Usually, a single vaccination does not make a person immune to an illness for life. Therefore, immunizations are performed over a period of years, or booster shots are needed periodically.

COOL BODY BITS

## A cure for smallpox

The word *vaccine* comes from the Latin word *vacca,* which means cow. Why? Because Edward Jenner first used a virus that affected cows — cowpox — to prevent a virus that affected people — smallpox — from causing so much illness. In 1798, Jenner took material from cowpox lesions (yes, from cows) and put it into people (inoculated them) so that the people would create antibodies against a pox virus. These antibodies then fought the even more dangerous and extremely contagious smallpox virus, which was an epidemic at the time. However, giving live viruses to people doesn't work for every illness; that's infecting them, not inoculating them.

## Passive immunity

You're born with passive immunity. When you were in your mother's womb, you received her antibodies as her blood coursed through your umbilical cord. Breastfeeding delivers additional antibodies that were secreted into your mother's breast milk. A few months out of the womb, and some of those original antibodies wore out. Your body had to start making its own, and your immune system has been functioning ever since.

# What Can Go Wrong with Your Immune System

The fine line from healthy to diseased can be crossed when a problem arises in the immune system. Some immune system disorders are evident immediately; others exist for years before any signs or symptoms occur. The three main types of immune system disorders are autoimmune diseases, allergies, and immunodeficiencies.

## Autoimmune diseases

These diseases occur when your immune system attacks your own "self" cells. Some bacteria and viruses produce toxins that cause T cells to attack the body's own proteins on the surface of the macrophage instead of the microbe's antigen that's inside the macrophage. Eventually, killer T cells start to view other cells in the body as foreign. Some diseases thought to be initiated in this manner include lupus erythematosus, rheumatoid arthritis, multiple sclerosis, and myasthenia gravis.

### Lupus erythematosus

Lupus affects connective tissue, causing an arthritislike condition. The two major types of lupus are *Discoid lupus erythematosus* (DLE), which affects connective tissues in the skin, and *systemic lupus erythematosus* (SLE), which affects several of the body's systems. Because SLE is much more common, this section focuses on SLE, which waxes and wanes through periods of attack (usually during spring and summer) and times of remission. The disease affects mostly women. An infection with a streptococcal bacterium or a virus, as well as pregnancy, ultraviolet light, stress, and faulty estrogen metabolism can cause lupus.

People with SLE experience flulike symptoms and pain in several joints *(polyarthralgia)*. Chest pain, difficulty breathing, low blood pressure, and tachycardia (rapid heart beat) occur in about half of SLE patients. Signs that neurologic damage has occurred include seizures, depression, irritability, headaches, and mood swings. Urinary tract infections and kidney failure are the most common causes of death in people with SLE. There is no cure for lupus. Treatment often includes corticosteroid (see Chapter 8) medications such as prednisone and medications to prevent problems in the kidneys and blood vessels.

### Rheumatoid arthritis

Inflammation caused by a person's immune system attacking its own cells eventually damages cartilage, then joints. This type of damage occurs in people with rheumatoid arthritis. Rheumatoid refers to the condition of rheumatism, which consists of inflammation, degeneration, and limited movement of structures made of connective tissue (such as joints and tendons). Flulike symptoms and development of joint inflammation ranging from swelling to destruction of bone, followed by atrophy, deformities, and loss of mobility, signal the disease. It usually begins in the fingers, but also can occur in other joints of the arms and legs, including the ankles, knees, wrists, and elbows.

The predisposition to developing rheumatoid arthritis runs in families (including mine). Hormone levels also may affect the development of the disease. Rheumatoid arthritis affects mostly women, starting between the ages of 35 and 50. There is no cure. Treatment consists of anti-inflammatory medications and pain relievers. Immunosuppressants (drugs that suppress the immune system to prevent the body from attacking its own cells) can be used early in the disease. Exercise and physical therapy help to retain a range of motion in the joints. As the disease progresses, surgery often becomes necessary to realign bones and reduce pain. Replacement of joints may be needed.

# Allergies

Allergens cause the immune system to have a hypersensitivity reaction. If certain allergens, such as pet dander, dust, or pollen, make you itch, break out in hives, have a runny nose (rhinitis), or make your eyes water, then you've experienced a hypersensitivity reaction. Allergies can go through cycles. Some allergies are seasonal, such as hay fever, whereas others occur year-round (such as food allergies). The severity of allergies can change from year to year, some allergies stop after a certain amount of time, and some start late in life without prior problems with a substance. Genetics can predispose people to allergies.

IgE antibodies, which cause the release of histamine, initiate allergic reactions. (Refer to the "Antibody-mediated immunity" section earlier in this chapter for more on IgE antibodies.) Histamine causes swelling of mucous membranes, such as in the nose and throat. The swelling causes nasal congestion and that annoying itch in the throat that you can't scratch. Congestion and swelling can trap bacteria in the nasal cavities and lead to sinus infections or ear infections.

A severe allergic reaction that causes sudden breathing difficulty is *anaphylaxis.* These reactions can lead to shock (anaphylactic shock) or death. People with severe allergies to foods such as shellfish or peanuts can experience these frightening reactions. Treatment is an epinephrine injection. Tracheotomy (creating a hole through the throat and trachea so that air can enter) sometimes is necessary if the tissues in the throat swell, obstructing airflow. Tissues lining the bronchial tubes of the lungs also can swell, obstructing air flow through the lungs. Respiratory failure, shock, and arrhythmias of the heart (altered rhythm of heart beat) can lead to death rapidly, so knowing the signs of anaphylaxis and acting quickly can save someone's life if their immune system begins to cross that fine line between health and disease.

# HIV and AIDS

AIDS is caused by the *human immunodeficiency virus* (HIV). HIV targets helper T cells, which makes a person infected with HIV unable to fight against this virus — a real Catch-22. Over time, the immune system of an HIV-positive person becomes deficient in helper T cells, creating an immunodeficiency (a deficiency of a part of the immune system). This immunodeficiency is acquired (rather than induced) through lifestyle choices (such as unprotected sex, intravenous drug use) or events (such as blood transfusion) that

expose a person to HIV. The presence of HIV and the deficiency of helper
T cells that develops as a result of infection by the virus usually causes an
affected person to acquire diseases such as pneumonia.

After having HIV in their body for a period of time — and becoming
immunodeficient — people usually develop the disease called *acquired
immunodeficiency syndrome* (AIDS). When a person progresses from being
HIV-positive to actually having AIDS, they tend to become infected with
bacteria and viruses easily, and cancer also can occur (such as Kaposi's
sarcoma). A cure for AIDS hasn't been found yet, and treatment is expensive
and still mostly experimental.

# Part IV
# Creating New Bodies

The 5th Wave    By Rich Tennant

OB GYN

RADIOLOGY →

"I may not know anything about genetics, but I know that I'm overdue and it's your side of the family that's always late for family functions."

## In this part . . .

Individual humans can survive without reproducing, and this is one reason why the information about the reproductive system has its own part. The human species can't survive, however, if individuals don't reproduce; so each of us has a strong natural urge to mate and reproduce. Details about the reproductive system, which contains the parts of the body that allow reproduction to occur, appear in Chapter 14. Chapter 15 describes how humans develop in the womb, what happens during the birth process, and how people change during their lifetime.

# Chapter 14

# What a Production! Reproduction

**R**eproduction is essential to continuing the human species, and the urge to reproduce is as instinctual now as the *fight-or-flight response* used to be. Humans are animals, and all animals know how and when to mate. Bulls don't sit their young offspring down and have "the talk." "Now, make sure that you find a nice cow, one who can provide plenty of milk for the calves." Mother cows don't explain the whole process to their young female calves, either. Animals mate with the strongest and fittest of the species. They don't know that it ensures the health of the species, but they don't agonize over relationships, either. Only humans, as smart as they are, have the need to understand their partner along with the process of mating and reproduction. So to satisfy your human curiosity, I've detailed how the reproductive systems of humans work. I'll leave the dating and mating rituals for another book.

# Going Forth from Gonads to Genitals

To reproduce — that is, to produce offspring — a male must fertilize a female. This is true not only for humans but for all animals and plants — yes, plants — see *Biology For Dummies* by Donna Rae Siegfried (Hungry Minds, Inc.). In mammals (that includes humans), the males deposit sperm in females to cause fertilization. Sperm are male sex cells called *gametes*. The sperm seek to fertilize an egg, which is the female gamete. Gametes are produced in the *gonads* (primary sex organs), which in males are the testes and in females are the ovaries.

*Genitals,* however, are the secondary sex organs used during sexual relations to transport the male gametes to the female gametes so that fertilization can occur. Male gametes are transferred to females through the genitals; sperm leave the penis (the male genital) of the male and are deposited in the vagina (the female genital) of the female. (You've got to get all these "G" words straight before you get going on the "S" word: sex.)

# Ladies First: The Female Reproductive System

The reproductive system of a human female is shown in Figure 14-1. The female's role in reproduction is to produce the eggs for fertilization and to transport the fertilized eggs through the uterine tubes to the uterus, where they're nourished and allowed to develop until birth (see Chapter 15).

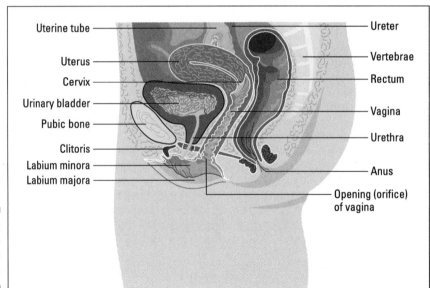

Uterine tube
Uterus
Cervix
Urinary bladder
Pubic bone
Clitoris
Labium minora
Labium majora

Ureter
Vertebrae
Rectum
Vagina
Urethra
Anus
Opening (orifice) of vagina

**Figure 14-1:**
The female reproductive system.

*From LifeART®, Super Anatomy 1, © 2002, Lippincott Williams & Wilkins*

The major players in the female reproductive system include the

✔ **Ovaries:** These two almond-shaped primary sex organs, one on each side of the pelvic cavity, are responsible for producing eggs. They're approximately five centimeters wide, and they house groups of cells called *follicles,* each of which is multilayered. The cell layers in the follicle surround an immature egg, called an *oocyte.*

✔ **Uterus:** The uterus (also known as the *womb*) is a muscular organ shaped like an upside down pear. Positioned over the bladder, the uterus is connected to the *uterine tubes* (also known as *fallopian tubes*). The walls of the uterus are thick and capable of stretching as a fetus develops. The uterus, which normally is 2 inches wide can stretch to 12 inches wide as a baby grows. The lining of the uterus, called the *endometrium,* becomes part of the placenta during pregnancy.

✔ **External genitalia:** In females, the external genitalia carries the term *vulva,* which refers to the labia majora, labia minora, and the clitoris. (See the section "The keys to your vulva" later in this chapter.)

✔ **Vagina:** The vagina secretes fluids and serves as a passageway for sperm to follow into the uterus.

✔ **Fallopian (uterine) tubes:** These tubes serve to transport the female gametes to the uterus.

All the female reproductive organs lie on the *broad ligament,* which is a large sheet of tissue that supports the organs.

All the eggs that a female will ever have are produced before she is even born. The oocytes remain in the ovaries throughout childhood and then start to mature and to be released (the process of *ovulation*) when puberty begins. Puberty is the period of great developmental change that results in a mature reproductive system. For females, puberty begins around age 11 and lasts until age 13. Several changes involving an increase in the sex hormone *estrogen* herald female puberty: the development of pubic and axillary (armpit) hair, the enlargement and development of breasts, and the beginning of the menstrual cycle. Estrogen, along with some other hormones, controls the menstrual cycle (see the next section, "Demystifying menstruation"). When menstruation begins, so does a female's ability to become pregnant and produce offspring. (See the "Creating a Pregnancy" section later in this chapter for more on becoming pregnant.)

## Demystifying menstruation

*Menstruation* is the period of time during the menstrual cycle (see Figure 14-2) in which menstrual bleeding occurs; therefore, it's known as a "period." But menstruation, which occurs for an average of 5 days, is only part of the *menstrual cycle,* which takes 28 or so days to complete. The menstrual cycle consists of both the *ovarian cycle* and the *uterine cycle,* which run concurrently to prepare both an egg and the uterus for possible pregnancy. Menstruation occurs when the levels of the hormones estrogen and progesterone decline; however, the entire menstrual cycle — that is, ovarian and uterine cycles — is directed by several hormones, not just estrogen and progesterone.

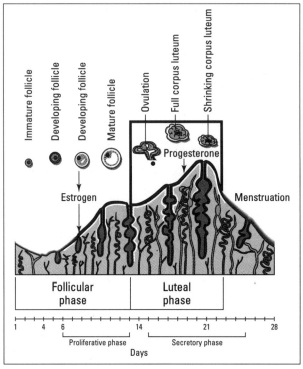

Immature follicle
Developing follicle
Developing follicle
Mature follicle
Ovulation
Full corpus luteum
Shrinking corpus luteum

Progesterone

Estrogen

Menstruation

Follicular phase

Luteal phase

1  4  6          14          21          28

Proliferative phase          Secretory phase

Days

**Figure 14-2:** The menstrual cycle.

*From LifeART®, Super Anatomy 3, © 2002, Lippincott Williams & Wilkins*

### *Looking at the ovarian cycle*

The ovarian cycle is the most important part of the menstrual cycle because it is responsible for producing the hormones that then control the uterine cycle. (See the "Clocking the uterine cycle" section coming up later in this chapter.) For the *first 13 days* (of a 28-day cycle), a hormone stimulates the maturation of an oocyte in one of the follicles within one of the ovaries. The hormone is appropriately called *follicle-stimulating hormone* (FSH), and the anterior part of the pituitary gland, which is in the brain, secretes FSH (see Chapter 8).

FSH is secreted when the hypothalamus (also in the brain and discussed further in Chapter 8) detects a low level of estrogen. The low level of estrogen signifies that FSH needs to be secreted so that the follicle develops. When the follicle is developed enough, it begins to secrete estrogen. When the level of estrogen reaches the appropriate level, the hypothalamus — as the master gland it is — uses feedback inhibition to "instruct" the pituitary gland to stop secreting FSH. When FSH secretion stops, the oocyte is fully mature and ready to be released. On *day 14* (of a 28-day cycle), ovulation occurs, and the oocyte is released from the ovary.

## PMS: Is it all in your head?

Well, kind of. *Premenstrual syndrome* (PMS) is a term used to describe the mood swings, irritability, emotional reactions, fatigue, food cravings, bloating, and cramping that occurs just before menstruation. Some women never experience these symptoms, some women have the symptoms mildly, and some women — well, let's just say that *everybody* knows when they're about to get their period! But what controls when a woman's period occurs? Hormone levels. And what controls hormone levels? Two little glands in the brain: the hypothalamus and the pituitary gland. The hypothalamus detects the levels of hormones and "tells" the pituitary gland when it should secrete hormones such as estrogen, progesterone, FSH, and LH. So although menstruation itself is occurring in the reproductive system at one end of the body, the other end of the body — the brain in the head — plays a big role in the ups and downs of hormone levels.

At the time of ovulation, the anterior pituitary gland, which has been secreting FSH and *luteinizing hormone* (LH) simultaneously, causes a surge in the secretion of LH. This LH surge causes the follicle from which the oocyte was released to become a *corpus luteum* (yellow body). The corpus luteum secretes the hormone *progesterone,* which triggers the hypothalamus. When the corpus luteum has secreted a sufficient amount of progesterone, the hypothalamus stops the anterior pituitary gland from secreting any more LH. At that point, the corpus luteum begins to shrink (about day 17). When the corpus luteum is gone (about day 26), the levels of estrogen and progesterone are at the lowest levels of the cycle (sometimes causing symptoms of premenstrual syndrome), and menstruation starts (about day 28).

Like any cycle, the whole process starts over. When the level of estrogen is low during menstruation, the hypothalamus detects the low level and secretes *gonadotropin-releasing hormone* (GnRH), which prompts the anterior pituitary gland to release its gonadotropic hormone — FSH — so that another follicle is stimulated to develop a new oocyte that secretes estrogen. Now, you're back to the first paragraph of this section.

### Clocking the uterine cycle

The 28-day uterine cycle, which aims to prepare the uterus for a possible pregnancy, overlaps with the ovarian cycle.

- ✔ **Days 1 to 5:** The *first 5 days* of the uterine cycle is when the level of estrogen and progesterone are lowest — the "period" of menstruation. The low level of sex hormones fails to prevent the tissues lining the uterus (the *endometrium*) from disintegrating and shedding. As the tissues tear apart from the wall of the uterus, blood vessels rupture, causing the bleeding that occurs during a period. The blood and tissue passes out of the uterus through the cervix and then out of the body through the vagina.

## Leaving the womb through the wrong exit

The endometrium is the lining of the uterus that's shed during menstruation. *Endometriosis* is the medical term that describes the condition of endometrial tissue growing in or on organs of the body other than the uterus. Usually, endometriosis is confined to organs in the abdominal cavity. How does the endometrial tissue spread to other organs? Well, during menstruation, some of the disintegrated tissue from the endometrium that's being discharged backs up through the uterine tube instead of passing out the uterus through the cervix. When an ovary releases an oocyte, an opening at the end of the uterine tube normally accepts the oocyte. That opening is the wrong exit for disintegrated endometrium.

The endometrial tissue is sensitive to the amount of hormones traveling in the bloodstream, whether the tissue is in the uterus where it belongs or in some other location. At the end of the uterine cycle, when the hormone levels decline, endometrial tissue disintegrates — no matter where it is. In women who don't have endometriosis, the shedding of the endometrium can cause cramps and pain. But when the disintegration is occurring at spots on other organs (such as the bladder, ovary, or large intestine), the pain can be extreme.

Surgery is necessary both to diagnose and treat endometriosis. Following removal of the endometriosis, it's necessary to stop the production of estrogen, so that the uterine cycle no longer continues. Drug therapy or removing the ovaries can stop estrogen production. Women with short menstrual cycles (fewer than 27 days) or long menstrual periods (longer than 7 days), tend to have an increased risk of endometriosis. Women who have taken oral contraceptives (birth control pills) for a long time or who have had several pregnancies have a lower risk of developing this condition.

✔ **Days 6 to 14:** During this *proliferative phase* is when estrogen production is highest. The developed follicle secretes estrogen, which makes the endometrium regenerate fresh tissue. The tissues lining the uterus and the glands in the uterine wall grow and develop an increased supply of blood. All these changes are preparation for nourishing an embryo and supporting a pregnancy, should the oocyte, which is released on day 14, become fertilized and implant in the wall of the uterus. (See the section "Creating a Pregnancy" later in this chapter.)

✔ **Days 15 to 28:** During this *secretory phase,* the corpus luteum secretes an increasing level of progesterone, which further thickens the endometrium, and the glands of the uterus secrete a thick mucus. If the egg becomes fertilized, the thickened endometrium and mucus help to "trap" the fertilized egg so it implants properly in the uterus. If the egg does not become fertilized within a day or two, the corpus luteum begins to shrink because it won't be needed for a pregnancy. As the corpus luteum shrinks, the progesterone and estrogen levels decline, which causes the endometrium to "shred and shed" just before menstruation.

Early on in a pregnancy, the corpus luteum serves as a source of progesterone until the placenta develops and can secrete progesterone on its own.

## Invaginating the vagina

*Invagination* describes the folding in of a surface to form a cuplike shape or a sheath, such as the vagina — the part of the female anatomy where the male penis penetrates a female's body during sexual intercourse.

The vagina, which is 7.5 to 10 centimeters long, is a tube that's open at the bottom. At the top of the vagina lies the cervix, which is the narrow, bottom end of the uterus. The cervix is a round, muscular structure that normally is open ever so slightly to allow sperm to pass into the uterus. During child-birth, the cervix opens wide to allow the fetus to move out of the uterus. The vagina (also called the *birth canal*) must accommodate a fetus that is being born, so the walls of the vagina are made of stretchy tissues: some fibrous, some muscular, and some erectile. In its normal state, the walls of the vagina have many folds, much like the lining of the stomach. When the vagina needs to stretch, the folds flatten out, providing more surface area.

The vagina does have a function besides being a receptacle for semen and a passageway for fetuses making their way into the world. When a woman is sexually aroused, the erectile tissues in her vagina fill with blood. The pressure of the increased volume in the blood vessels there forces fluid out of the walls of the blood vessels. This fluid lubricates the vagina, making it easier and more comfortable for a penis to penetrate. The fluid also helps to transport sperm through the cervix and into the uterus, so that fertilization can occur.

## The keys to your vulva

No, a vulva is not a practical, boxy sedan. *Vulva* is the collective term for the female's external genitalia: the clitoris, labia majora, and labia minora. The term *labia* means *lip*, and the labia of the vulva are loose flaps of flesh, just like the lips of the mouth (called *labia mandibulare* and *labia maxillare*, by the way). The labia protect the opening to the vagina and cover the bony structures of the pelvis.

✔ **Labia majora:** These large folds of skin — one fold on each side — cover the smaller labia minora. The labia majora extend from the *mons pubis* (pubic mound) back toward the anus. The mons pubis contains fat deposits that cover the pubic bone. Following puberty, pubic hair covers the mons pubis and the labia majora.

✔ **Labia minora:** These hairless folds of skin lie underneath the labia majora and cover the opening of the vagina. The labia minora are attached near the vaginal orifice (opening) and extend upward, forming the foreskin that covers the clitoris.

✔ **Clitoris:** This part of the vulva located above the opening to the vagina and above the urethra, has a shaft and glans tip, just as a penis does, and it's extremely sensitive to sexual stimulation. The clitoris contains erectile tissues that fill with blood during sexual stimulation. Because the tissue of the labia minora cap the clitoris, the swelling and reddening is also obvious in the labia minora.

Stimulation of the clitoris can lead to orgasm in the female. Although females don't ejaculate, females do experience a building and release of muscular tension. Female orgasm causes the muscle tissue that lines the vagina and uterus to contract, which helps to pull the sperm up through the reproductive tract.

# Following the Ladies: The Male Reproductive System

Males are responsible for producing sperm and for transmitting sperm to a female so that fertilization of an egg can occur. Of course, those have always been their biological responsibilities regarding reproduction. The evolution of human society has given males more responsibilities in the way of caring for and helping their partners, as well as in parenting and supporting offspring. Biologically, though, the male contribution to reproduction hasn't changed much over the years.

The internal anatomic structures of the male reproductive system (see Figure 14-3) include the testes, the gonads where sperm are produced, and vas deferens, which are tubes that carry sperm from the testes to the penis. The external anatomic structures of the male reproductive system include the penis and scrotum. Several tubes and glands are involved in the production and dissemination of sperm. (See more on these terms later in this section.)

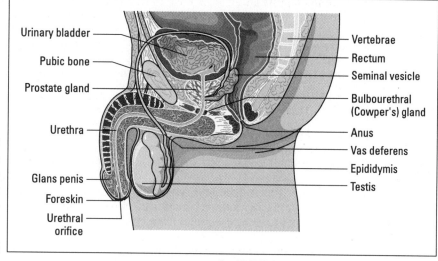

*From LifeART®, Super Anatomy 1, © 2002, Lippincott Williams & Wilkins*

**Figure 14-3:**
The male reproductive anatomy, sagittal view.

Labels: Urinary bladder, Pubic bone, Prostate gland, Urethra, Glans penis, Foreskin, Urethral orifice, Vertebrae, Rectum, Seminal vesicle, Bulbourethral (Cowper's) gland, Anus, Vas deferens, Epididymis, Testis

# Inside the male reproductive system

The *testes* (plural) are held with one of the external male structures, the scrotum. Each testis (singular) is an egg-shaped gland that produces sperm and hormones. The testes are akin to the ovaries of the female in that they produce the gametes.

The testes are covered with and contain fibrous tissue, creating compartments within each testis. Inside the compartments are the five feet of coiled *seminiferous tubules,* which produce *spermatozoa* (*sperm* for short) during a process called *spermatogenesis* (the process of sperm development and maturation). The seminiferous tubules form the *epididymis,* which lies on top of each testis. The epididymis, like the seminiferous tubules, is a long cordlike structure that ends in a coil on top of a testis (refer to Figure 14-3). The epididymides (plural) serves as a "holding tank" for maturing sperm. Each epididymis is continuous with ("continuous with" means "becomes") a *vas deferens,* which is the final place of maturation for sperm. The vas deferens is a tube that connects the epidydimis of each testis to the penis.

The walls of seminiferous tubules are lined with thousands of *spermatogonia* (developing sperm). The seminiferous tubules also contain specialized cells — the Sertoli cells — that provide nourishment for the developing sperm and that regulate how many of the spermatogonia are developing at any one time.

The spermatogonia go through mitosis (see Chapter 2) and produce another generation of spermatogonia. Those new spermatogonia move into the lumen (the hollow part) of the tubule so that they have room to grow. When they are big enough, they are then called *primary spermatocytes.*

The primary spermatocytes go through meiosis (see Chapter 2) so that when they divide, each new spermatocyte has half of the chromosomes, some having an X sex chromosome and some having a Y sex chromosome.

Humans have 46 chromosomes, but half (23) come from the mother's egg, and the other half come from the father's sperm. Meiosis in primary spermatocytes is how sperm come to have 23 chromosomes.

The spermatocytes with 23 chromosomes go through another cell division, producing four *spermatids,* each having 23 chromosomes. The spermatids become *spermatozoa* when the head and tail have fully formed. A mature sperm is composed of a head, a short body (or middle piece) and a long tail called a *flagellum.* The head contains the 23 chromosomes inside its nucleus; the head also is covered by a structure called an *acrosome,* which contains enzymes that break down the membrane of the egg to allow the sperm to bore into the egg and fertilize it. The middle part of the sperm contains *mitochondria,* which are organelles that produce the energy compound that the body uses for fuel: ATP (see Chapter 2). A sperm needs ATP to provide energy to move the flagellum, which allows the sperm to "swim."

As opposed to females, who produce their gametes through a cycle, the process of spermatogenesis occurs continuously in males. Males are born with spermatogonia in their seminiferous tubules, but they (the spermatogonia, that is) don't begin to mature until puberty. During puberty, the hypothalamus releases gonadotropin-releasing factor, and it stimulates the anterior pituitary gland to secrete follicle-stimulating hormone (FSH). Yes, although males don't have egg-containing follicles, they do have FSH. Perhaps a better name for that hormone would be "gamete-stimulating hormone." After the male brain begins secreting FSH, it doesn't stop (as it does during the monthly ovarian cycle of a female).

Males also produce luteinizing hormone (LH), although they don't create a corpus luteum like a female does. (For more on LH, refer to the "Looking at the ovarian cycle" section earlier in this chapter.) In males, the LH works in a feedback loop with testosterone. *Testosterone,* a sex hormone, is produced in *interstitial cells* (also called *Leydig cells*), which occupy spaces among the seminiferous tubules inside the testes (*interstitial* means between parts or in the spaces between cells in a tissue). When testosterone levels decline, more LH is secreted. The rising LH level spurs on the production of testosterone, and when testosterone comes back up to normal level, the secretion of LH ceases.

Testosterone may be considered the male hormone, but women also produce testosterone in small amounts. Estrogen masks testosterone, so when a woman's estrogen level declines (such as after menopause), the effects of testosterone in her system can be seen (like the growth of facial hair).

Mature sperm can live in the epididymis and vas deferens for up to 6 weeks. If sexual activity occurs within that time, the sperm are added to seminal fluid (semen) that's produced in the male. Semen is a milky liquid produced when several glands in the male reproductive tract secrete fluid.

1. The *seminal vesicles,* glands located at the juncture of the bladder and vas deferens, have ducts that allow the fluid that they produce to sweep the sperm from the vas deferens into the urethra.

2. Next, the *prostate gland* adds its fluid, which contains mainly citric acid and a variety of enzymes that keep semen liquefied. The prostate gland surrounds the urethra just below where the duct of the seminal vesicle joins the urethra.

   When the prostate gland swells (enlarges), it squeezes the urethra, which makes urination difficult and sometimes painful. Enlarged prostate glands usually occur in older men. The condition is so common that it might seem to be a normal part of the aging process, but in fact an enlarged prostate gland is abnormal.

3. The *bulbourethral* (or Cowper's) *glands* sit just below the prostate gland on either side of the urethra. They're two small glands that have ducts leading directly to the urethra.

These three types of glands — the seminal vesicles, the prostate gland, and the bulbourethral glands — secrete fluids that have several functions:

- ✔ They're slightly basic with a pH of 7.5, just the way sperm like their environment to be.

- ✔ They nourish the sperm by providing the sugar fructose so that the sperm's mitochondria can make enough energy to move its tail and travel all the way to the egg.

- ✔ They contain *prostaglandins,* which are chemicals that make the uterus contract. When the uterus contracts, the sperm are pulled further upward into the female's reproductive tract.

As the secretions are added from the glands, forming the semen, pressure builds up on the structures of the male reproductive tract. When the pressure has reached its peak, the semen is expelled out of the urethra through the penis. Peristaltic waves (like those that occur in the digestive tract, see Chapter 11) and rhythmic contractions move the sperm through the vas deferens and urethra. The term for this discharge is *ejaculation* — part of orgasm in males, as is the contraction and relaxation of skeletal muscles at the base of the penis. As the muscles contract rhythmically, the semen comes out in spurts.

Each ejaculation produces about one teaspoon of semen, which contains about 400,000,000 sperm.

## Outside the male reproductive system

The external genitalia of males consists of the scrotum, which holds the testes, and the penis, the organ of sexual intercourse.

The *scrotum* (a pouch of skin) holds the testes outside of the body because sperm thrive better in a cooler temperature. Body temperature is several degrees too warm for sperm to remain viable. During fetal development (see Chapter 15), the testes in males are located inside the abdominal cavity, but just before birth (and sometimes shortly after birth), the testes descend into the scrotum. If they don't descend, the condition is *cryptorchidism,* and surgery is usually required for correction. If left untreated, viable sperm can't be produced, and the male is unable to produce offspring.

The scrotum also contains smooth muscle, which contracts when the scrotal skin senses cold temperatures (such as when swimming), and pulls the scrotum up closer to the body to keep the sperm at the right temperature. The inside muscle layers of the scrotum are an outpouching of the abdominal cavity. The outside skin of the scrotum is continuous with the skin of the perineum and groin.

Quickly stroking the skin on the upper thigh with something like a popsicle stick causes the scrotum to move the testes closer to the body in what is called the *cremasteric reflex.*

The penis is, of course, the star of the male reproductive anatomy. However, it's also part of the male urinary system. Both the urinary and reproductive systems share the urethra as the tube that carries both urine and semen out of the body. However, semen doesn't contain urine. When ejaculation is about to occur, a sphincter closes off the bladder to keep urine, which is acidic, from mixing with the sperm, which live in a basic environment.

The penis consists of a shaft and the *glans penis* (tip). The urethra runs through the penis; the glans penis contains the urethral orifice. *Foreskin* (also called *prepuce*) covers the glans penis. Foreskin is often removed shortly after birth in a minor surgical technique called *circumcision.*

During sexual arousal, the penis becomes erect. When the male is stimulated sexually, the brain sends impulses through the parasympathetic nervous system (see Chapter 7) that cause dilation of the arteries in the penis. As the arteries dilate, the veins are compressed. So as blood flows into the penis, it doesn't flow out, building pressure. This pressure forces blood to flow into the sinuses of erectile tissue. As the erectile tissue fills with blood, the penis hardens. Erection enables the penis to be inserted into the female so that intercourse and thus reproduction can occur.

*Impotence* is the condition where erection doesn't occur following sexual stimulation. Impotence has a variety of possible causes, including damaged blood vessels (sometimes due to diseases such as diabetes), psychological factors (stress, fear), or nerve damage. Treatment depends on the cause, of course, but can include drug therapy, surgery, counseling, or mechanical devices or implants that can create an erection.

# Creating a Pregnancy

Do not confuse "getting pregnant" with "becoming fertilized." They aren't one in the same. In order for a pregnancy to occur, several things must happen:

1. Sexual intercourse between a man and a woman must take place, resulting in ejaculation of semen from the man's penis into the vagina of the woman.

2. One sperm from the man's semen must successfully travel from the vagina, through the cervix, up through the uterus, and into the uterine tube where an oocyte that has just been released from its follicle in the ovary is being held.

3. The sperm that makes it to the oocyte must be able to digest away the material on the outside of the oocyte (the *corona radiata* and *zona pellucida*) and get inside the egg to its nucleus.

4. The sperm must join with the nucleus of the oocyte, combining the genetic material of the man with the genetic material of the woman.

If the sperm makes it to Step 4, fertilization has occurred. But *IF* is a really big word because many complicating factors exist. Timing is everything. If intercourse and ejaculation occur too far prior to ovulation, the sperm will be dead before the egg is released — sperm live for 12 to 48 (maybe even 72) hours inside the female. If intercourse and ejaculation occur too far after ovulation, the egg disintegrates by the time the sperm arrive — oocytes live for only 12 to 24 hours after ovulation.

Even if the timing is right, that doesn't mean that conditions are suitable for the sperm. A woman's vagina is a pretty acidic place, so most of the sperm, which thrive in a base environment, are killed before they even make it to the cervix. If it's not the day of ovulation, the mucus in the cervix can be pretty thick, which prevents more of the sperm from passing through. On the day of ovulation, the cervical mucus is watery, thus allowing more sperm to enter the uterus. The chance for fertilization is better.

If the conditions and timing are right, and fertilization occurs, that still doesn't mean the woman is pregnant. More work needs to be done.

A fertilized egg merges 23 male chromosomes with 23 female chromosomes, creating a cell with 46 chromosomes. At this point, the cell is a *zygote*. The zygote goes through several cell divisions as it moves slowly down the uterine tube to the uterus. If the zygote makes it to the uterus and properly embeds itself into the endometrium, then pregnancy has occurred. The woman has conceived. The movement of a zygote to the uterine lining takes several days.

After the zygote has implanted in the lining of the uterus, it's an *embryo*. To keep the pregnancy going, the embryo must begin to produce a hormone called *human chorionic gonadotropin* (hCG). The hCG prevents menstruation, and thus prevents disintegration and discharge of the lining of the uterus (as well as the embryo).

If conception takes place but the embryo doesn't produce hCG, the embryo and the uterine lining is discharged as if it were menstruation. If this process happens before the woman even knows she was pregnant, she may just consider it a late, heavy period. If it happens after she realizes she's pregnant, it's considered a *spontaneous abortion (miscarriage)*.

The hCG usually can be detected within 10 to 14 days after the time menstruation should have started. At this point, the woman may have experienced tender breasts, morning sickness, and probably has some idea that she's pregnant. Urine pregnancy tests detect the presence of hCG. If hCG is present, a pregnancy is confirmed.

# Preventing a Pregnancy

The time that's right for fertilization to occur may not necessarily be right for a woman or a couple to get pregnant. Whether it's a *good time* or a *bad time* to engage in sexual intercourse becomes a matter of opinion, and people who don't want to become pregnant have a variety contraception methods (birth control) that they can use to prevent fertilization (conception) from occurring.

Nothing, of course, beats *abstinence*. When you don't want to become pregnant, avoiding sexual intercourse is the most fail-proof method of birth control. However, abstinence doesn't always make the heart grow fonder, so couples who desire sexual intercourse but not pregnancy have several options.

# For the gentlemen

*Condoms* for males are rubber sheaths worn on the penis to catch the ejaculate and prevent sperm from entering the vagina. Males also have the option of undergoing a *vasectomy,* which is a minor surgical procedure in which the vas deferens are severed and then tied off, preventing any sperm that are produced in the testes from ever being included in the semen. The male also can withdraw his penis from the female prior to ejaculation, a method known as *coitus interruptus.* However, because live sperm commonly seep out prior to withdrawal and ejaculation, this method isn't totally effective

# For the ladies

Females have many more contraceptive options. *Oral contraceptives* (birth control pills) prevent the anterior pituitary gland from releasing FSH and LH, thus preventing oocytes from maturing and ever being released — no ovulation, no pregnancy. Oral contraceptives can either provide estrogen and progesterone or just one of those hormones. Without ovulation, no secretion of LH — neither estrogen nor progesterone — occurs; therefore, with low estrogen and progesterone levels, menstruation nevertheless still occurs.

Injections of Depo-Provera also alter hormone levels and thus suppress ovulation. Intramuscular injections of this contraceptive are given every three months. Altering the hormone levels is an effective but not risk-free means of preventing pregnancy. Blood clots can occur in women taking oral contraceptives, and breast cancer or osteoporosis can occur in women who receive Depo-Provera shots.

Morning-after pills are drugs that cause the termination of a newly established pregnancy or prevent pregnancy from occurring. These medications are options for women who have unprotected sexual intercourse but then realize that they may have been ovulating at the time and don't want to become pregnant. In other words, they don't have to wait to see whether pregnancy occurs and then consider surgical abortion or adoption.

Morning-after pills are not limited to *mifepristone* (RU-486), which sparked considerable debate because it was available for years in France and Great Britain but not in the United States. In September 2000, mifepristone was also approved for use in the U.S. Other morning-after pills have been approved for several years. After taking a morning-after pill to interrupt a pregnancy, a woman must also take a prostaglandin pill, which causes contractions that expel the contents of the uterus. A couple of risks involved with using morning-after pills are heavy bleeding that must be stopped surgically and incomplete emptying of the uterus, which requires a surgical abortion.

Females also can undergo a *tubal ligation* (surgically cutting and tying the fallopian tubes) when they desire permanent sterility. When the uterine tubes are tied, released eggs aren't in the proper position for fertilization, or they cannot travel to the uterus.

*Spermicides,* or chemicals that kill sperm, come in various forms: foams, creams, and jellies. They can be inserted into the vagina prior to intercourse to kill sperm before they can travel higher into the reproductive tract. Unfortunately, spermicides alone are only about 75 percent effective, possibly because they may not always be applied for complete coverage of the top of the vagina and yet sperm can be deposited there.

Some female contraceptives work as *barriers.* The *diaphragm* is a latex cup that covers the cervix and prevents sperm from passing into the uterus. The *cervical cap,* made of rubber, is held onto the cervix by suction. The *intrauterine device* (IUD) is coiled plastic that is placed in the uterus; with such an object already in the uterus, a zygote won't implant itself in the uterine wall. These objects also carry a risk of problems, such as pelvic inflammatory disease, infection, or possible cervical cancer.

*Female condoms* also are available. Made of latex sheathing and inserted into the vagina, they serve as a barrier that prevents sperm from getting through the cervix.

# Pathophysiology of the Reproductive System

The reproductive system is one of the few systems of the body that has contact with other people or objects (the digestive system is another). Therefore, bacteria and viruses can easily be spread to other people via the organs and secretions of the reproductive system. In addition, structural defects, hormonal problems, and genetic abnormalities can cause problems in the reproductive system. And the reproductive system is no stranger to cancer.

# Sexually transmitted diseases

Just as their name says, *sexually transmitted diseases* (STDs) are transmitted via sexual contact, and they affect both males and females. Condoms can limit the spread of many STDs, and people who are unsure if their partner has an STD or HIV may use them. (See the "AIDS" section later in this chapter.) The following sections cover some of the most common STDs.

## Genital warts

The *human papillomaviruses* (HPVs) cause *genital warts*. Genital warts are bumps that appear on the external genitalia of males or females. However, sometimes genital warts are flat, so they go unnoticed, which means a person might not even know he or she is infected with this virus. And, an unknowing person can unknowingly spread this STD. Genital warts can be removed, but they often recur. Tumors and cancer are a consequence of contracting genital warts. The HPVs that cause genital warts also may be responsible for nearly all cases of cervical cancer, and they are associated with tumors on the vulva, in the vagina, and on the anus or penis.

## Genital herpes

*Herpes simplex virus type 2* (HSV-2) causes genital herpes. Herpes simplex virus type 1 (HSV-1) is the type that causes cold sores and fever blisters. A person with infected with HSV-2 may never have symptoms, but symptoms can include painful, recurring ulcers that form on the penis or vulva. When the ulcers recur, the person may develop a fever, swollen lymph nodes, and experience painful urination. If a pregnant woman is infected with HSV-2, cesarean section usually is performed when the infant is ready to be born. Otherwise, passing through the birth canal can expose the infant to a serious infection that can cause brain damage or death.

## Chlamydia

*Chlamydia* causes infection of the urethra similar to gonorrhea. (For more on gonorrhea, see the following "Gonorrhea" section.) Because of the similarity, physicians often treat both gonorrhea and chlamydia at the same time. But because the bacteria *Chlamydia trachomatis* is the culprit in chlamydia, the condition it causes is *nongonococcal urethritis* (NGU). If a woman is infected with chlamydia during the time that she's giving birth, the bacteria may get into the newborn's eyes, causing inflammation, or in the newborn's lungs, causing pneumonia.

In addition to affecting the urethra, chlamydia can cause ulcers to form on the cervix. If the infection goes undetected for a long period of time, it can lead to a condition called *pelvic inflammatory disease* (PID), which can occur in males or females. In males, PID causes the vas deferens to become inflamed. The tubes swell from the inflammation, and as they heal, scar tissue forms, which can become partially or completely blocked, leading to infertility. In women, the same process of inflammation and healing occurs in the uterine tubes. The scar tissue can block the tubes, preventing an egg from becoming fertilized or, if the egg does become fertilized, preventing it from implanting in the uterus. (See information on ectopic pregnancy in Chapter 15.)

### Gonorrhea

*Gonorrhea* generates symptoms similar to that of chlamydia, but a different bacterium, *Neisseria gonorrhoeae,* causes gonorrhea. Gonorrhea causes infection of the urethra, which produces a thick, yellow-green discharge and painful urination.

Gonorrhea may rear its ugly head after the initial infection as pelvic inflammatory disease. Then the vas deferens of a male or the uterine tubes of a female may become inflamed, heal, and scar. The scar tissue can block the tubes, causing infertility or sterility. An infant being born through a birth canal affected with gonorrhea usually develops an eye infection that can cause blindness. For this reason, all newborns are administered antibiotic ointment immediately after delivery.

### Syphilis

The bacteria called *Treponema pallidum* causes syphilis, and it develops in three stages.

1. During the primary stage, an ulcer with hardened edges appears on the genitals. The ulcer heals, and time goes by in which the infected person has no symptoms.

2. In the second stage, a rash covers the entire body, including the palms of the hands and soles of the feet. The rash heals, and more time passes without symptoms.

3. By the third time that symptoms appear, the disease is serious. The person may have cardiovascular effects or nervous system effects, such as blindness, mental retardation or insanity, and a shuffling gait. At this time, ulcers called *gummas* can develop on the skin or in the organs.

   A woman with syphilis who becomes pregnant can pass the bacteria to her fetus because *T. pallidum* can cross the placenta. If the fetus becomes infected, it can develop birth defects or die before birth (stillbirth).

### AIDS

*AIDS,* acquired immunodeficiency syndrome (see Chapter 13), is spread through sexual contact, so it can be considered an STD. AIDS is the result of infection with the human immunodeficiency virus (HIV), just as syphilis is the result of infection with *T. pallidum.*

# *Infertility*

*Infertility* is the inability to fertilize or be fertilized. Saying that a couple is infertile means that they've been trying to create a pregnancy for at least one year but haven't been successful. *Sterility* is the total inability to have children.

Blocked uterine tubes (such as from scarring or endometriosis) and failure to ovulate are the most common causes of infertility, but it's not always the woman who is infertile. Men can produce too few sperm or too many abnormal sperm. Abnormal sperm occur regularly, even in men who have fathered children. Some sperm have two heads, two tails, or are missing pieces. However, as long as fewer than 25 percent of the sperm are defective, infertility usually isn't an issue. A man's ejaculation normally contains about 400 million sperm, and 100 million abnormal sperm may seem like too many. But that leaves 300 million normal sperm that could fertilize the egg.

Men with testicular disease often have low testosterone levels. The genetic disorder *Klinefelter's syndrome* (XXY) causes men to have an extra X chromosome. (Women normally have XX; men have XY.) As children, the boys develop normally, but when they get to puberty, the testes fail. Without testosterone being produced in the Leydig cells of the testes, a wide variety of effects can be seen: development of breast tissue, impotence, bone loss, and infertility.

Previously healthy men who experience trauma to the testicles or aging men who experience a decline in the function of their Leydig cells also can become infertile. And certain bacterial or viral infections, such as mumps, can result in *orchitis* (inflammation of the testes) that can lead to *hypogonadism* (decreased function of the testes or ovaries) — both of which can affect fertility.

Pituitary tumors in males or females also can cause hypogonadism by virtue of the fact that the pituitary gland secretes FSH, which normally spurs on maturation of the oocyte or spermatocyte and the subsequent release of estrogen or testosterone. Problems with the pituitary gland can cause symptoms of hypogonadism, which in women includes the loss of menstruation (amenorrhea) and infertility. In men, the symptoms of hypogonadism are impotence and infertility.

The cause of infertility must be determined before treatment can begin. But treatment options (and hope) for couples who desire a pregnancy are available.

# Cancer

In men, the most common types of cancer in the reproductive system are prostate cancer and testicular cancer. Women are most commonly affected with ovarian cancer or cervical cancer. (See Chapter 15 for more on breast cancer.)

## Prostate cancer

The prostate gland surrounds the urethra, so a tumor in the prostate gland can compress the urethra, making urination difficult, more frequent, and painful. Prostate cancer can occur during middle age, but it most commonly affects elderly men. If the flow of urine is completely blocked, urine backs up into the bladder and ureters. Kidney problems can result. The cancer also can spread up through the bladder and ureters to the kidneys. Prostate cancer also can lead to bone cancer.

Physicians check for enlargement of the prostate gland through the rectum. If the prostate gland feels hard, ultrasound and biopsy are performed. Treatment involves removing the prostate gland and radiation to prevent more cancerous growth. In addition, to prevent the spread of cancer, the testosterone level must be kept low, which is done by removing the testes or administering estrogen.

## Testicular cancer

Cancerous growth of tissue in a testis can affect the seminiferous tubules and Leydig cells, resulting in decreased production of sperm and testosterone, both of which can cause infertility. (See "Inside the male reproductive system" section earlier in this chapter for more on seminiferous tubules and Leydig cells.) Sterility is possible if both testes are affected.

The typical method of diagnosis and treatment is removal of the affected testis, followed by radiation or chemotherapy (drugs). If the other testis is unaffected, fertility usually can be preserved. Although testicular cancer is much more rare than breast cancer, men are encouraged to perform self-examinations of their testes to check for lumps, just as women are encouraged to perform self-examinations of their breasts.

## Ovarian cancer

Cancer of the ovary is most common in women older than 50 years. Preventing pregnancy and experiencing pregnancy both seem to protect against ovarian cancer because this type of cancer is two to three times more common in women who never bore children and less common in women who took oral contraceptives. Ovarian cancer can begin in the ovary or appear in the ovary after spreading from another structure (usually the breast).

Ovarian cancer doesn't usually cause symptoms until it's widespread. The woman may experience abdominal pain and swelling, nausea and vomiting, gas, or fluid in the abdominal cavity. Abnormal vaginal bleeding is rare. *Laparoscopy* is a minor surgical procedure in which the physician can view the organs in the abdominal cavity.

If the physician finds cancer, the affected ovaries can be removed during the same laparoscopic procedure. However, major surgery may be necessary: A *salpingo-oophorectomy* is removal of the fallopian tubes (*salping-*) and ovaries (*ooph-*); a *hysterectomy* is removal of the uterus. Radiation and chemotherapy follow the surgery to prevent cancerous growth from recurring.

If ovarian cancer affects just one of the ovaries, the survival rate is 60 to 70 percent within five-years. (Meaning that 60 to 70 percent of women will be alive after five years.) However, if both ovaries are affected, the five-year survival rate drops to 10 to 20 percent. Pap smears don't detect ovarian cancer, so women older than 40 years are encouraged to get a "cancer check" every year during their regular gynecologic or physical examination.

### Cervical cancer

The *Pap smear* (short for Papanicolaou, the last name of the physician who created the test) detects abnormal cells on the cervix 95 percent of the time. Abnormal cervical cells can lead to cervical cancer, which can be cured completely if it's diagnosed early enough. Cervical cancer can easily spread to other organs in the pelvis and then through the lymph nodes. Therefore, detection and follow-up of abnormal cells (*dysplasia*) is important.

Thanks to Dr. Papanicolaou's standard test — which women are advised to undergo once a year — abnormal cells usually are detected in women before any symptoms develop. If abnormal cells are detected, a *colposcopy* (inspection of the cervical tissue) is performed. However, the early sign of cervical cancer is abnormal vaginal bleeding (that is, nowhere near the time of the menstrual period). If the cancer spreads into other pelvic organs, the woman experiences pain.

If precancerous cells are detected while a woman is pregnant, treatment is postponed until the infant is delivered. Cervical cancer that's detected and treated early carries a 50 to 80 percent five-year survival rate. When the cancer has spread to other organs and radiation treatments are necessary, the survival rate can drop to 10 to 30 percent.

Cervical cancer is strongly linked to the human papillomavirus. Smoking, having many sexual partners, or starting to have sex at an early age are risk factors for cervical cancer. (For more on the human papillomavirus, refer to the "Genital warts" section earlier in this chapter.)

# Chapter 15

# Here I Grow Again: Birth and Development

*L*ife is like a story that's in the process of being written. It starts with a preface — fertilization and fetal development. Chapter 1 begins with your birth. Each stage of life adds a new chapter with constant editing — changes taking place. The story ends with death at the end of the final chapter. It doesn't sound like a happy ending, but you don't know it yet. We're still writing our stories. Perhaps an epilogue can explain that at the end of our story: You look back in amazement and wonder at the story that unfolded.

## Tackling Trimesters

"First comes love, then comes marriage, then comes baby in a baby carriage . . . " The old schoolyard song doesn't really apply to every couple. Marriage and even love aren't requirements for mating and reproducing. Sperm don't check for the marriage license before they seek the egg. The process of fertilization doesn't require any amorous feelings. Sperm meets egg, egg becomes fertilized, egg implants in uterus, pregnancy achieved. And you thought the fun was over!

Many changes occur during pregnancy. Most changes obviously affect the *fetus* (unborn child from the third month until birth), but the expectant mother also experiences many changes in her body, too. This section explains the changes that occur in both mother-to-be and fetus throughout the 40 weeks of pregnancy, which are broken up into three 13-week trimesters.

If you want even more information on pregnancy than I can provide you in one chapter, check out *Pregnancy For Dummies* by Joanne Stone, MD, Keith Eddleman, MD, and Mary Murray (Hungry Minds, Inc.) for a whole book full.

## Developing cleavage: The first trimester

The first trimester technically begins before you even have sexual intercourse. Usually by the time a woman discovers that she's pregnant, she's already gone through 4 to 6 weeks of the 40 weeks, which is longer than the 9 months that's commonly referred to as pregnancy's time frame. The 40 weeks of pregnancy are counted from the beginning of your last menstrual period. Why? Because ovulation isn't always on the 14th day of the ovarian cycle, fertilization doesn't always occur on the same day as intercourse, and the amount of time it takes for the fertilized egg to implant varies (see Chapter 14). The date of the last menstrual period is a point from which everyone can start. From that day, you go forward 280 days to determine the due date for a fetus to be born. Keep in mind, though, that because of all the variables and uncertainties regarding exact timing, only five percent of babies are actually born on their due dates. (Only one of my three children was born on the actual due date.)

Below is a step-by-step guide to the highlights of human development throughout the three trimesters of pregnancy.

1. **Pregnancy occurs when the *zygote*, the fertilized egg that has gone through several cell divisions, finds a nice comfy spot in the uterine wall and implants itself in the tissue.** (See Chapter 2 for more on cell division.)

2. **The implanted *embryo*, as the former zygote is now called, makes its home in the *endometrium* (the lining of the uterus) until it's time to be born.** (For more on the endometrium and the uterus, see Chapter 14.)

3. **The embryo begins producing *human chorionic gonadotropin* (hCG), which is the pregnancy hormone.**

   During the first trimester, the mother-to-be experiences several changes in her breasts: They enlarge and become tender as the glandular tissue inside grows to become ready for producing milk, the area around the nipple (the *areola*) darkens, and veins in the breasts become more prominent. Fatigue is no stranger, either; pregnant women are often fatigued during the first trimester as the body tries to adjust to all the changes going on inside her. Urination becomes more frequent as the growing uterus presses in on the bladder, and "morning" sickness can occur in the afternoon and evening as well. The presence of hCG in the mother-to-be's system makes some women nauseous.

   At this point, the woman may consider a home-pregnancy test or may call her physician for a pregnancy test. Either way, her urine is tested for the presence of hCG. If hCG is in the urine, then, yep, she's going to have a baby.

4. **During the embryonic stage (the first eight weeks of pregnancy), the placenta and amniotic sac develop.**

   **Placenta:** This is a special organ formed in the uterus during pregnancy that serves as the place where the mother-to-be's blood deposits nutrients and oxygen, and the fetal blood picks them up. The fetal blood is carried through the umbilical cord, which connects the fetus to the placenta. Then the wastes that result from the fetus metabolizing the nutrients and oxygen are carried back out through the umbilical cord and deposited into the placenta. The mother-to-be's blood picks up the wastes from the placenta, and her body excretes them. Geez, moms start cleaning up after their kids before they're even born!

   **Amniotic sac:** This is like a water balloon in which the developing embryo floats. The sac is formed from two thin membranes — the *chorion* and the *amnion* — and the fluid is *amniotic fluid*. Just as water holds up an aboveground pool, the amniotic fluid keeps the membranes from pressing against the developing embryo/fetus so that growth isn't hindered. The fluid also keeps the temperature constant for the fetus and absorbs the shock from the mother's movements.

   All animals, even those that live on dry land, develop in a water environment, such as amniotic fluid.

5. **The organs begin developing during the first month of pregnancy.**

   All the body's organs, including the brain and spinal cord, are formed by the end of the first month of pregnancy. The circulatory system forms from small vessels in the placenta called *chorionic villi* just three weeks after fertilization occurs. The heart begins to beat and doesn't stop until the moment of death, hopefully many decades later.

# Amniocentesis

Amniocentesis, as in "having an amnio," is a procedure in which a long, thin needle is inserted into the abdomen and into the uterus. Some of the amniotic fluid is withdrawn into the needle. Because the fetus is floating and moving around in the amniotic fluid, fetal cells are found in the clear fluid. Genetic screening can be performed on those cells to check for certain abnormalities, such as Down syndrome, or the presence of certain infections can be ruled in or out.

The procedure is performed between 15 and 20 weeks although not all pregnant women have it done. Women who earlier had abnormal results on an ultrasound or a screening blood test usually undergo amniocentesis. Late in pregnancy, amniocentesis can be performed to assess if the lungs of the fetus are mature enough for delivery. The lungs produce chemicals that are in the amniotic fluid that keep them from sticking together. If the fluid contains those beneficial chemicals, then the fetus is mature enough for pre-term delivery. Following amniocentesis, amniotic fluid may leak out for a day or two until the membranes re-grow and seal the puncture site.

6. **During the second month, the organ systems continue to develop, and the limbs, fingers, and toes begin to form.**

7. **The embryo begins to move at the end of the second month.**

   It's still too small for the mother-to-be to feel its movements.

8. **Ears, eyes, and genitalia appear during the second month.**

9. **The embryo loses its tail and begins to look less like a seahorse and more like a human.**

   You do more developing when you're an embryo than at any other time in your life. Several developmental processes occur during the embryonic stage:

   **Cleavage:** The term for one cell dividing into two through the process of mitosis (see Chapter 2). After the sperm combines its genetic material with the egg, the now-called zygote goes through series of cell divisions, doubling the number of cells over and over. The entire zygote does not get larger as this happens; instead, the cells just increase like soap bubbles in a confined space. The number of cells increase, but the size of each of them decreases. Still, each cell contains all the genetic material that was in that first original cell.

   **Morphogenesis:** The movement of cells to a different place in the blastocyst is *morphogenesis* (meaning "creation of shape.") As the zygote goes through divisions and moves down the uterine tube prior to implantation in the uterus, it becomes a *morula* and then a *blastocyst*. The cells inside the blastocyst rearrange themselves so that two layers of cells form. The inner layer *(inner cell mass)* becomes the embryo upon implantation. The outer layer *(trophoblast)* becomes the chorion from which the placenta (and then the circulatory system as well as other body systems) is created.

   **Differentiation:** After the cells take shape during morphogenesis, they begin to develop specific structures and functions. To do this, they need to become different types of cells; the process that the cells go through to do this is differentiation, and genes control it (see Chapter 2). When differentiation begins, the cells become specialized.

   The specialized cells must "commit" to forming certain types of tissues, or they lose their ability to do so. These shiny new cells — *stem cells* — have the capability of becoming almost any type of cell. Stem cell research is so promising because of the large potential and important uses for cells that can differentiate into a wide variety of specialized cells. The specialized cells form certain types of tissues, and similar tissue types get together to form organs and organ systems.

## Chorionic villus sampling

Chorionic villus sampling (CVS) is another procedure that yields some valuable information. Between the tenth and twelfth weeks of pregnancy, some placental tissue is withdrawn through a needle inserted into the abdomen or through a catheter placed in the cervix. The tissue contains cells from the chorionic villi, which have developed from the fertilized egg just as the fetus has. By testing the chorionic villi cells, the fetus and its amniotic sac is left undisturbed. The cells can be cultured within seven to ten days and can provide information as to whether any chromosome abnormalities and certain diseases are present, and what the gender of the fetus is (if gender is a factor in acquiring or carrying a familial disease). This information allows women and their mates to determine whether to continue an affected pregnancy.

10. **Around 8 weeks, the embryonic period is over.**

   From the third month until birth, the baby is called a *fetus*.

At the end of the first trimester, the fetus is about 4 inches long and weighs one ounce. The head is large, and hair begins to grow on it. The intestines are inside the abdomen, and the urinary system (kidneys and bladder) start to work with the fetus producing urine.

# A womb without a view: The second trimester

Everything gets better in the second trimester. The mother-to-be's body has adjusted to the increased demands on her system, so she experiences less fatigue. And the level of hCG declines as the placenta is now fully functioning to support and nourish the fetus, so the nausea that the mother-to-be experiences declines, too.

1. **The fetus — with all of its systems in place — grows.**

   As it grows, the mother-to-be notices the movements that the fetus makes — a kick to the abdomen here, an elbow to the ribs there, or a somersault in its amniotic sac. The fetus sucks its thumb and opens its eyelids. It acts like a newborn, but it's still wrinkled with see-through skin; fine, downy hair called *lanugo;* and a greasy, white material (picture solid shortening smeared all over the fetus) called *vernix caseosa.* The gender of the fetus as well as the skeleton can be seen on ultrasound during the second trimester.

2. **Bone begins to replace the cartilage that formed during the embryonic stage.**

At the end of the second trimester, the fetus is about 12 to 14 inches in length and weighs about three pounds. It probably wouldn't survive if born at this point, but it's getting close.

For the mother-to-be, the second trimester is totally opposite from the first trimester. Although minor annoyances, such as heartburn, flatulence, and congestion, can begin during the second trimester, the mother-to-be usually has more energy and feels great throughout these months. Her hair and nails grow thicker, and her skin glows due to the increased level of hormones flowing through her body. The mother-to-be accepts her pregnancy, she looks pregnant, and her excitement for the birth builds.

## Seeing the light: The third trimester

The last trimester of pregnancy is an emotional time. Apprehensiveness about giving birth may develop, and the discomforts that result from the growing fetus can make the mother-to-be a bit, well, let's say "cranky." (Just ask my husband!)

The fetus grows fast and furiously during the last trimester. (A third-trimester fetus is shown in Figure 15-1.)

1. **Fat is deposited under its skin, which makes it look more like a baby and less like an old man.**

   The deposition of fat helps to insulate the fetus at birth, but it also increases the weight of the fetus and serves as a critical energy reserve for brain and nervous system development.

2. **The fetus, with its systems developed, continues to grow in size.**

## Seeing double

Occasionally, the inner-cell mass separates, creating two or more embryos. Because the embryos develop from cells that contain exactly the same genetic material, they're called *identical twins*. *Fraternal twins* develop from two different eggs that are fertilized by two different sperm; basically, fraternal twins are like any other siblings except that they were in the womb at the same time.

**3. Near the end of the third trimester, the fetus positions itself for birth.**

The mother experiences the fetus dropping lower into the pelvis (called *lightening*), as the uterus descends and takes pressure off the diaphragm. The fetus normally turns head down and aims for the exit. When the head of the fetus reaches the ischial spines of the pelvic bones (see Chapter 4), the fetus is said to be *engaged* for birth.

The mother-to-be usually gains about 28 pounds throughout the pregnancy, but most of it is gained during the last trimester. During the last few weeks of the pregnancy, the uterus has reached maximum capacity, expanding from its normal size of 2 by 3 inches to fill the abdominal cavity all the way up to the ribs. The size of the expanded uterus and the pressure of the full-grown fetus may press on the diaphragm and make it difficult for the mother-to-be to breathe. Sleeping can be difficult, too; not only because of the discomfort from the uterus pressing on the ribs but also because of the uterus pressing on the bladder, creating the urge to urinate several times per night. A woman in the third trimester of pregnancy can also experience a tearing sensation on the sides of her abdomen, which results from the stretching of the round ligaments that support the uterus. Expectant mothers everywhere lodge other complaints: swelling, back pain, itchiness, varicose veins, carpal tunnel syndrome, and clumsiness are just a few more.

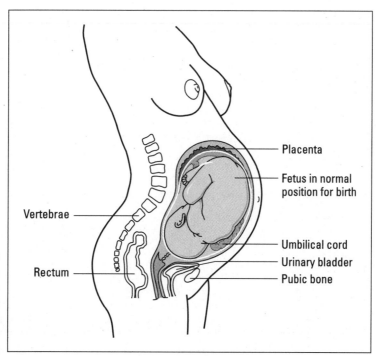

**Figure 15-1:** Development of the fetus, third trimester.

Placenta
Fetus in normal position for birth
Umbilical cord
Urinary bladder
Pubic bone
Vertebrae
Rectum

*From LifeART®, Super Anatomy 3, © 2002, Lippincott Williams & Wilkins*

Whatever the discomforts, whatever the pains, they soon come to an end, as the time to deliver the fetus is near. By the end of the third trimester — when the fetus is ready to be delivered — it's about 21 inches long and 8 pounds.

# Complicating Things

Unfortunately, sometimes trouble besieges a pregnancy. When you consider how many things can actually go wrong during a pregnancy, it's amazing that so many trouble-free births occur. See the following list of some of the problems that can occur during pregnancy.

If you're pregnant while reading this, don't be alarmed. If you think you have any of these problems, let your physician know of your concerns.

- **Bleeding:** Spotting can occur early in the first trimester. When the zygote embeds in the uterus, some blood can be released. This blood may appear around the time the woman may be expecting her period, so it may cause some alarm if she already knows she's pregnant. However, it's normal. Unless abdominal cramping accompanies the bleeding or is bright red, a miscarriage probably hasn't occurred. Physicians should be told of any unexpected bleeding, though. Later in the pregnancy, the pressure of the heavy fetus and large uterus can cause hemorrhoids in the rectum, which can bleed occasionally. Although this bleeding is a pain in the butt for the mother-to-be, it poses no harm to the fetus.

- **Ectopic pregnancy:** An ectopic pregnancy occurs when the fertilized egg embeds outside of the uterus, such as in the uterine tube. The pregnancy takes hold as it normally would in the uterus with the embryo producing hCG and developing organs. The problem is that the uterus is the only place where room is available for an embryo to grow. If the zygote implants in the uterine tube (or cervix or ovary), abdominal pain and bleeding usually result. A dangerous complication is that the tube (or ovary or cervix) can rupture as the embryo grows. Ultrasound can detect ectopic pregnancies. Because they're a threat to the mother-to-be's health, they must be removed, as they cannot be transplanted to the uterus.

- **Gestational diabetes:** Pregnant women are tested for the presence of glucose in their urine at every prenatal visit. Late in the second trimester, a glucose screen is often performed to help catch developing cases of gestational diabetes. Gestational diabetes is hyperglycemia (too much glucose in the blood) that develops during pregnancy. If too much glucose is in your blood, then too much glucose is in the blood of the fetus as well. Too much glucose can cause the fetus to grow too large, making delivery difficult and sometimes necessitating a cesarean section. The development of gestational diabetes can lead to diabetes mellitus after the pregnancy. If glucose levels can't be kept down through a careful diet and exercise, insulin may need to be administered to the mother-to-be.

✔ **Incompetent cervix:** If the cervix begins to dilate and open long before the fetus is ready to come out, a miscarriage has a good chance. The mother-to-be may feel a heavy pressure in her pelvis, but the uterus doesn't contract to open the cervix. The cervix is just unable — that is, incompetent — to support the pregnancy. Women who've had procedures performed that required dilation of the cervix or inflicted trauma on the cervix (such as biopsy) are at a higher risk of having an incompetent cervix. Incompetent cervix also may affect women who have torn their cervix during a prior delivery or who are carrying three or more fetuses at the same time. To alleviate the problem, in a procedure called *cerclage,* the cervix can be stitched to give it extra support.

✔ **Pre-eclampsia and eclampsia:** Just as urine is checked for glucose to help stave off gestational diabetes, the urine is also checked for protein to help prevent pre-eclampsia from developing. Pre-eclampsia — the condition that can lead to the more severe eclampsia — is the development of high blood pressure during pregnancy. Severe headache, sudden swelling, rapid weight gain (five pounds in one week), blurry vision, and abdominal pain are all symptoms. The danger of pre-eclampsia is that the high blood pressure can lead to eclampsia (seizures and possibly coma or even death). If the mother-to-be's health is at risk, and the fetus would survive if delivered, a cesarean section is performed to deliver the baby.

✔ **Placenta previa:** In this condition, the placenta covers the cervix, partially or completely, preventing delivery of the fetus or causing heavy bleeding. If placenta previa happens early in pregnancy, it usually fixes itself because the placenta moves up into the uterus as the fetus grows. (Remember that the placenta is attached to the umbilical cord.) If it happens in the second or third trimester, however, the blood-filled placenta may begin to leak blood as the growing fetus presses on it. Bleeding can stimulate preterm labor, which would result in a premature infant. Most people with placenta previa end up having a cesarean section to prevent heavy bleeding that could occur during a vaginal delivery.

✔ **Placental abruption:** This condition occurs when the placenta separates from the wall of the uterus before the fetus is ready to be delivered. It can cause bleeding, and the bleeding can bring on preterm labor. If the tear is large, the fetus needs to be delivered immediately. If the tear is small, bed rest may be tried, and the pregnancy is followed closely.

# The Labor of Love

Okay. It's time. Time to explain what happens to the human female when a fetus is ready to be born. No matter what a woman goes through to deliver a child, the effort is rewarded. The first reward is holding the newborn for the first time, and staring at your new baby completely amazed at what the human body can do. From there, rewards are plentiful: the baby wrapping its tiny hand around your finger, the smell and feel of a newborn on your skin (even if it's 3 a.m.), smiles and giggles, and hearing "I Love You, Mama" for the first — well, every time. Continuing the species, creating life, feeling connected to the past and the future — those are the rewards of giving birth. Women just usually aren't thinking about them while they're in labor.

In this section, you take a look at the birth process: the three stages of labor and delivery as well as what happens right after giving birth. To delivery and beyond!

## Moving down: Stage one

Before the fetus can be delivered, it must move down through the birth canal. This process happens during stage one of labor (see Figure 15-2). When true contractions start (the ones that dilate the cervix rather than just tighten up the uterus; those "warm-ups" are called Braxton-Hicks contractions), labor starts, and stage one begins. In addition to the beginning of regular, effective contractions, other signs can let the woman know that "it's time." However, this first stage can last 12 to 14 hours or more! So if it's "time," take your time and stay relaxed.

Some women notice that their mucous plug has been released. During pregnancy, a thick glob of mucus fills the opening of the cervix to prevent any sperm or bacteria from getting inside the uterus. After the cervix begins to dilate, the plug falls out, and a bit of blood can accompany it. Sometimes the loss of the mucous plug is the "bloody show." However, not all women notice that they've lost this mucus; they may think it's just vaginal discharge, which is pretty commonplace late in pregnancy.

**Figure 15-2:**
Early labor and active labor: Dilation and effacement of the cervix, transition phase.

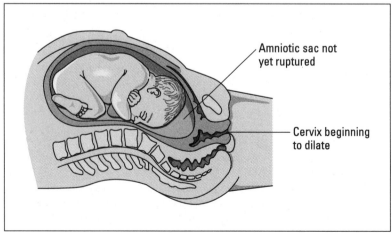

Amniotic sac not yet ruptured

Cervix beginning to dilate

*From LifeART®, Super Anatomy 3, © 2002, Lippincott Williams & Wilkins*

As the cervix is dilating, the head of the fetus continues to press on the cervix. The pressure causes the cervix to thin out and move up or efface. Effacement and dilation are necessary for the fetus to be delivered. The cervix must open ten centimeters (about 4 inches) to allow for the diameter of the head to pass through. Effacement is necessary so that the cervix becomes thin like a pancake as it widens; normally, the cervix is closed and about an inch long and rests at the bottom of the uterus like a thimble. It cannot widen without flattening (100 percent effacement).

As effacement occurs, more of the amniotic membrane is exposed through the cervix. The pressure of the contractions pushing the fetus down on the membranes can cause the membranes to rupture, which allows amniotic fluid to leak (or gush) out — the "water breaks." Losing amniotic fluid usually kicks contractions into high gear and if they weren't regular before, they become regular.

By the end of the first stage of labor, the cervix is 100 percent effaced and dilated about 4 centimeters. Contractions occur regularly, about five minutes apart and lasting about one minute. Contractions aren't necessarily painful. The abdomen hardens as the uterine muscle contracts to push the baby down. The mother-to-be feels the tightening of the abdomen and the pressure of the baby pushing down. She's now in active labor.

During *active labor,* the contractions continue every three to five minutes and last 45 to 60 seconds. The pressure is building, and within about four to five hours, the cervix is dilated eight to nine centimeters. The fetus has moved down the birth canal and is just about ready to enter the world.

Women handle this stage differently; some want to walk, some want to lay down. Whatever position is most comfortable is important because what they really need to do is relax and save their energy for the "transition phase."

The *transition phase* — the time during which the fetus is born — can be considered the end of stage two of labor. During this phase, the cervix completes its dilation to the full ten centimeters. The contractions that widen the cervix the last centimeter or two are intense. But luckily, this phase is fairly quick. Now it's "time."

## Moving out: Stage two

After the cervix is completely dilated, the woman feels the urge to bear down and actively push (see Figure 15-3). All through labor, the contractions of the uterine muscles have been pushing the fetus down through the birth canal. Now, the mother-to-be can feel the fetus in the vagina and pressing on the pubic bones or rectum or passing the hip bones. The fetus is coming out, and the expectant mother's body has the urge to contract skeletal muscles to help get it out.

At this point, the woman is usually stopped from pushing so that the physician can make sure the cervix is completely dilated to prevent her from tearing it if it's not. I'm sure they mean well, but when you have to push, you have to push! The mother-to-be unintentionally grimaces, clenches fists, curls toes, and even pops blood vessels in her eyes; at that point, the fetus is undoubtedly coming through the cervix.

Pushing doesn't usually take long. Usually, within an hour after the urge to push strikes, the mother-to-be is holding her newborn baby. When it's time to push, the physician and nurses begin using the term *station.* As the fetus moves down through the vagina, the countdown is on. "Minus 1 station" means the fetus is about midway down the vagina. "Plus 2 station" means it's about 3-4 good pushes away. "Plus 5 station" means push and the head should come out.

Pushing is much like straining to have a bowel movement, except that you usually are curled up like a C laying on its side, pressing your chin to your chest, and having your legs back as far as they can go. The woman who is delivering may think to herself, "isn't this the position that got me to this point in the first place?" But it's more likely that she'll focus her energy, count to ten, and push the head of her fetus out of her vagina. At this point, the physician interrupts again, suctioning blood, mucus, and amniotic fluid out of the baby's mouth and nostrils. Suddenly, another push, and the body follows. The fetus is now an *infant* taking its first breath of life.

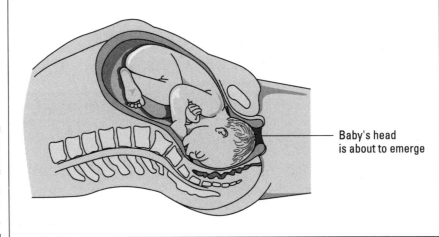

**Figure 15-3:**
Second
stage of
birth:
Delivering
the fetus.

Baby's head
is about to emerge

*From LifeART®, Super Anatomy 3, © 2002, Lippincott Williams & Wilkins*

## *Cleaning up: Stage three*

Yes, birth is messy. Along with the baby comes fluid and blood, and sometimes urine and feces from the mother. But don't start cleaning up the delivery room yet! Just after delivery, the infant's umbilical cord is cut and tied off. The infant is now totally separated from the mother and will soon have a stylish belly-button.

During its time in the womb, the infant — as embryo and fetus — relied on the placenta for antibodies, nourishment and exchange of wastes and carbon dioxide for nutrients and oxygen. The circulatory system and the intestines replace the placenta and umbilical cord. Now, the newborn infant is cut off from that former lifeline to the mother and must begin building its own immune system. Breathing, drinking, eating, urinating, and defecating must work on their own now.

After the baby is delivered (Figure 15-4), uterine contractions continue so that the placenta separates from the wall of the uterus. About 15 minutes after the baby is born, the placenta delivers. Even after the placenta (or afterbirth) comes out, uterine contractions continue. These contractions — which are most noticeable in the hours after delivery and then decrease over the next week or so — help the uterus to start returning to its normal size, and they force out excess blood and clots.

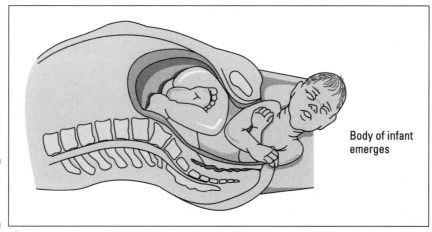

Body of infant emerges

**Figure 15-4:**
Emerging baby.

*From LifeART®, Super Anatomy 3, © 2002, Lippincott Williams & Wilkins*

## Putting away the push urge

Personally, I thought this was the hardest part of delivery. It's hard to stop pushing when the urge is so strong. And not letting the fetus continue its way out puts tremendous pressure on the rectum and perineum. Undergoing natural childbirth, I felt a burning sensation as I held back the urge to push, which probably was my skin tearing. It doesn't really take long to suction the airway, but it can sure seem like it when you're in the middle of delivery.

# *Making trouble from the start*

A trouble-free pregnancy doesn't always mean a trouble-free delivery, and a difficult pregnancy doesn't always mean a complicated delivery. Although problems can occur, the majority of full-term pregnancies produce a healthy child and cause no trouble for the mother. Two of the most common problems that crop up before and after delivery are

### *Fetal distress*

Upon arrival in the labor room (or birthing suite), the mother has an elastic band wrapped around her abdomen that's hooked up to a monitor. This device records the heart rate of the fetus and the occurrence and strength of contractions. If contractions cause the fetal heart rate to drop, and the heart rate does not come back up quickly enough, the fetus is said to be in "distress." That means that the fetus is not handling the stress of labor very well and may not be receiving enough oxygen.

A fetus in distress may need to be delivered more quickly than natural labor is allowing. A cesarean section may need to be the method of delivery, or other interventions may be necessary to help the mother through labor. Dehydration affects blood flow, which can cause fetal distress, so sometimes the mother needs extra fluids given intravenously to improve the situation. Or sometimes the mother just needs more oxygen, and after she breathes through an air mask, labor improves. And sometimes fetal distress is relieved just by the mother changing her position and moving around. Cesarean section isn't performed unless absolutely necessary.

### *Maternal hemorrhage*

Bleeding occurs after delivery, whether the delivery was vaginal or cesarean. The uterus continues to contract after delivery to help squeeze remaining blood out and close off blood vessels. Normally, bleeding stops about seven days after delivery. However, if the uterus doesn't contract after delivery, bleeding can be excessive. The condition is *uterine atony* (meaning "without tone," as in muscle tone), and it can occur when some placental tissue remains in the uterus after delivery, if an infection is in the uterus or if two or more fetuses were delivered.

To get the uterus to begin contracting again, a nurse may massage the abdomen over the uterus. Then a medication that causes uterine contraction, such as oxytocin, methergine, or Hemabate, may be given. If the cause of the uterine atony is remaining placental tissue, the tissue can be removed manually or by a dilation and curettage (D&C; casually referred to as a "dusting and cleaning").

In rare cases, the uterus can invert (turn inside out) upon delivery, and this condition causes hemorrhage, too.

# Developing Throughout Life: The Real "After Birth"

Birth is a beginning to a life filled with change. Just look how different people are in old age than they are in infancy. Changes and developments got them there.

## Starting fresh: Infancy and childhood

Infants grow and develop rapidly. In fact, many developments that began during the embryonic stage are still occurring. For one, the brain began its development during the third week of gestation, but it continues its development throughout the first several years of life. Then, even after it's completely developed, neural connections continue to be modified throughout life. The skeletal system also begins development during fetal life, but bone growth — that is, true lengthening of the bones — occurs until about age 20. And replacement of bone cells and tissues occurs throughout life. Although the organ systems are in place and working when a baby is born, the size of most organs increases until growth ends during the late teens and early 20s.

The reproductive organs of boys and girls are in a state of suspended animation until puberty occurs during the teen years. (See the following sections for more on puberty and the teen years.) Spermatocytes are present in the testes of boys, but sperm don't mature until puberty begins. Oocytes are present in the ovaries of females, but they don't become mature or released until puberty begins. Breast development doesn't occur in females until puberty. Male children have erections from the time they're born, but they're unable to reproduce until sperm mature. Muscle tissue increases during childhood, as well, which is why activity is so important for kids. Children between the ages of 7 and 12 should get one to two hours of vigorous activity every day.

## Pining away during puberty

"It was the best of times, it was the worst of times. . . . " I don't think Charles Dickens was referring to the teenage years when he wrote that line at the beginning of *A Tale of Two Cities*. But many people feel that way as they look back on their adolescent years. During adolescence, health is at a peak, but so are hormone levels.

Growth spurts are common during puberty and the teen years. Bones lengthen, muscle mass increases, and nearly adult size is reached. But the most important changes that occur during adolescence are the ones associated with development of the reproductive systems.

Puberty is the period of time when the reproductive systems of males and females matures to the point that reproduction can occur. In girls, puberty occurs between the ages of 11 and 13 years; in boys, it happens between the ages of 14 and 16. At this point in life, bone growth comes to an end, and adult height has been reached. When the epiphyseal plates at the ends of long bones (such as the femur) close, further growth can't occur (see Chapter 4). Around the end of the teen years, puberty is over, and the boy or girl has usually adjusted to the new hormone levels. However, the levels of sex hormones are at their peak, triggering the urge to mate.

The hormones produced during puberty can cause a period of adjustment for female and male bodies. The surge of hormones have been linked to triggering acne, and they often cause mood swings, as they affect the brain.

### Female puberty

In females, the ovarian and uterine cycles begin (see Chapter 14), which becomes evident when menstruation starts. After the female is ovulating, pregnancy is possible. The female *breast,* which is a gland (the mammary gland) that provides milk — and makes humans mammals — develops during puberty.

The breast contains about two dozen lobules, which are filled with alveoli and contribute to ducts. The ducts merge at the nipple. Inside the alveoli are milk-producing cells. During puberty, the lobules and ducts form, and adipose tissue is deposited under the skin to protect the lobules and ducts and give shape to the breast. During pregnancy, hormones increase the number of milk-producing cells and increase the size of the lobules and ducts.

After the infant is born, the pituitary gland secretes the hormone prolactin, which causes the milk-producing cells to create milk, and lactation begins. Lactation continues as long as a child nurses regularly. Suckling of the nipple stimulates nerves that end in the areola (the dark area around the nipple). That nerve impulse travels to the hypothalamus, which causes the pituitary gland to secrete oxytocin. Oxytocin is the hormone responsible not only for contracting the uterus during labor, but also for contracting the lobules in the breast so that milk flows out into the ducts. The infant suckles the milk out of the ducts through the nipple.

Other changes that occur in females during puberty include growth of pubic hair in the *axillary region* (armpit) and *pubic region,* and the development of the female fat distribution pattern: more on the hips, thighs, and breasts.

### Male puberty

Boys begin to develop sperm regularly during puberty, which causes them also to produce testosterone. Testosterone has certain effects, such as causing the growth of facial and chest hair, building lean muscle, causing pubic hair to develop in the axillary and groin regions, causing hair on arms and legs to become dark and coarse. The vocal cords lengthen, which causes deepening of the voice, and the penis and testes enlarge. Males develop broader shoulders and narrower hips than females. These characteristics of the male sex occur over the period of years during puberty.

## Growing up and getting creaky

Adulthood starts during the 20s and continues until death. Adults are fully grown and (usually) capable of reproducing. After growth is completed, caloric requirements decrease, and activity usually does, too. Therefore, adults must adjust to taking in fewer calories or expending more energy or they'll gain weight.

Over a period of years of consuming too many calories and being too inactive, the effects of extra weight can begin to affect the body's major systems. Table 15-1 lists some age-related changes to the body's systems.

| Table 15-1 | Age-Related Changes to the Body's Systems and Associated Health Implications | |
|---|---|---|
| **Body System** | **Change** | **Implications** |
| **Circulatory System** (see Chapter 9) | Heart increases in size | Increased risk of thrombosis and heart attack |
| | Fat is deposited in and around the heart muscle | Varicose veins develop |
| | Heart valves thicken and stiffen | |
| | Resting heart rate decreases | |
| | Maximum heart rate decreases | |
| | Pumping capacity declines | |
| | Arterial walls fill with plaque and don't stretch well | |
| **Digestive System** (see Chapter 11) | Loss of teeth | Increased risk of hiatal hernia, heartburn, peptic ulcers, constipation, hemorrhoids, and gallstones |
| | Peristalsis slows | |
| | Diverticulosis | |
| | Liver requires more time to metabolize alcohol and drugs | Colon cancer and pancreatic cancer increased in elderly |
| **Endocrine System** (see Chapter 8) | Glands shrink with age | Thyroid disorders and diabetes can occur |
| **Immune System** (see Chapter 13) | Thymus gland shrinks with age | Cancer risk increases |
| | Number and effectiveness of T lymphocytes decrease with age | Infections more common in elderly |
| | | Autoimmune diseases (such as arthritis) increase |

| Body System | Change | Implications |
|---|---|---|
| **Integumentary System** (see Chapter 6) | Epidermal cells are replaced less frequently | |
| | Fibers in the dermis thicken and have less collagen, are less elastic | Skin loosens and wrinkles |
| | Adipose tissue in face and hands decreases | Sensitivity to cold increases |
| | Fewer blood vessels and sweat glands | Body is less able to adjust to increased temperature |
| | Melanocytes decrease | Hair grays, skin becomes more pale |
| | Number of hair follicles decreases | Hair thins |
| **Muscular System** (see Chapter 5) | Mass and strength decrease | |
| | Muscle tissue deteriorates and becomes replaced by connective tissue or fat | |
| | Fewer mitochondria in muscle cells | Endurance decreases due to fewer mitochondria |
| | Cardiovascular and/or nervous system changes | Function can decrease due to cardiovascular or nervous system changes |
| **Nervous System** (see Chapter 7) | Brain cells die and are not replaced | Learning, memory, and reasoning decrease |
| | Cerebral cortex of brain shrinks | Reflexes slow |
| | Decreased production of neurotransmitters | Alzheimer's disease occurs in elderly people |

*(continued)*

## Table 15-1 *(continued)*

| Body System | Change | Implications |
|---|---|---|
| **Reproductive System** (see Chapter 14) | *Females:* menopause occurs between 45 and 55 years of age and causes cessation of ovarian and uterine cycles, so eggs are no longer released and hormones such as estrogen and progesterone are no longer produced | Wrinkling of skin, osteoporosis, and increased risk of heart attack |
| | *Males:* possible decline in testosterone level after age 50; enlarged prostate gland; decreased sperm production | Vascular (or other) problems can cause impotence in males |
| **Respiratory System** (see Chapter 10) | Breathing capacity declines | Risk of infections such as pneumonia increases |
| | Gas exchange and lung volume decreases due to thickened capillaries, loss of elasticity in muscles of rib cage | |
| **Skeletal System** (see Chapter 4) | Cartilage calcifies, becoming hard and brittle | Osteoporosis risk increases |
| | Bone resorption occurs faster than creation of new bone | More time is required for bones to heal if they break |
| **Urinary System** (see Chapter 12) | Kidney size and function decreases | Kidney stone risk increases |
| | Decreased bladder capacity | Incontinence |
| | Enlarged prostate gland in men | More frequent urination urges |
| | | Urinary tract infections more likely |

I know that's a pretty long list of changes that occur as a person ages. A healthy diet, exercising regularly, and drinking enough fluids can prevent many of those changes. Most of the problems encountered from middle age and beyond stem from poor lifestyle choices made early during the teen years or adulthood. Smoking, overeating, lack of exercise, lack of relaxation all contribute to serious problems later in life: heart disease, cancer, and diabetes are the three most common.

The development of a human being is an amazing process. Life is truly a gift and a source of wonder. Cherish the gift and take care of yourself so you can experience the possibilities of what your human body can do. See Chapter 16 for "Ten Ways to Keep Your Body Healthy" for some tips to turn into habits.

# Part V
# The Part of Tens

The 5th Wave                    By Rich Tennant

"Frank here used to teach high school physiology, so if you value your Zygomatic arch or your Alveolar margins, you'll start talking."

# In this part . . .

Okay, David Letterman doesn't have the market cornered on lists of ten. In this part, you get one of the great highlights of any *For Dummies* book, informative lists of ten things to know. Chapter 16 shows ten great ways to keep your body healthy — think of it as an owner's maintenance checklist. Chapter 17 offers up ten terrific Web sites that provide additional information about topics in this book and resources for your own Internet searches. The Web sites are pretty cool! I hope that you find them useful.

# Chapter 16

# Ten Ways to Keep Your Body Healthy

● ● ● ● ● ● ● ● ● ● ● ● ● ● ● ● ● ● ● ● ● ● ● ● ● ● ● ● ● ● ● ● ● ● ● ● ● ● ● ● ● ● ●

*In This Chapter*

▶ Getting what your body needs

▶ Watching out for trouble

● ● ● ● ● ● ● ● ● ● ● ● ● ● ● ● ● ● ● ● ● ● ● ● ● ● ● ● ● ● ● ● ● ● ● ● ● ● ● ● ● ● ●

*Y*ou take your car for oil changes every 3,000 miles; and you change the belts, fluids, and air filters regularly. You get tune-ups done on the car to make sure the machine that moves your body around town doesn't break down. But what about that other machine — your body? With proper care and maintenance, your body can outlast even the most reliable of cars. Following is a maintenance checklist of ten things that you can do to keep your own amazing machine in tip-top shape. The items in this list don't appear in "top ten" fashion — each one of these tips is important.

## Drink Water

Staying well hydrated is extremely important for your health, and there's nothing better than water to do the job. Regular coffee, tea, and colas contain caffeine, which acts as a diuretic. Although these drinks are wet, they don't contribute to keeping you hydrated. If you are thirsty, your body is telling you that it needs more water to keep tissues moist, blood flowing well, joints and waste moving, and metabolism running. Drinking something that contains caffeine sets you back further. But I can understand that drinking only water can get boring, so add lemon for a nice change, or drink decaffeinated beverages. Even eating soup or fruits can add water to your system.

Without enough water in your car's radiator, the engine can overheat. Right? The same thing can happen in your body. Sweat contains water, and your body releases sweat when it needs to let heat escape to cool down your body temperature. If you're dehydrated, your body is less able to regulate body temperature.

Water also plays a role in blood pressure. The higher the blood volume — that is, the more fluid there is in blood — the higher the blood pressure. Blood volume and blood pressure go hand in hand. So if you are dehydrated, your blood volume decreases, which makes your blood pressure decrease. This decrease may seem like a good thing, but there's more. When your blood pressure decreases, your arteries don't need to be open as wide to let a larger volume of blood through. The arteries constrict, increasing blood pressure and you get a headache. (And did you ever notice that it takes longer to have blood drawn when you are dehydrated?)

If the blood pressure is low in the arteries, it's even lower in the veins — therefore, waste remains in the body longer because it takes longer for the venous blood to return to the heart. Dehydration also affects the way your body releases waste: Your urine is more concentrated; that is, there's less of it, and it's darker yellow (more solutes, less solution). In addition, your stools are harder (constipation); and hemorrhoids can develop from straining or hard stools. Instead of releasing water through urine and feces, your body conserves it for all the important things that water does inside the body (like fueling your metabolic reactions).

Help your body maintain its balance — drink enough water to stay well hydrated. What's enough? At least eight 8-ounce glasses per day — that's 64 ounces, 2 quarts, or half a gallon, every day. (That's almost 2 liters per day, for you metric mavens.) Protect your body. Carry a water bottle with you at all times, and drink to your health.

# Eat Your Veggies (and Fruits), Dear

Fruits and vegetables. Vegetables and fruits. Eat more — 3, 5, or 9 servings every day is the recommended amount. Fruits and vegetables are simply good for you, and most people don't eat enough, which means you aren't getting the vitamins, minerals, and fiber that fruits and veggies provide. Instead, more people are taking multivitamin and mineral supplements or adding fiber to their orange juice. If you just eat the oranges instead, you'd be getting vitamins, minerals, and fiber all at the same time.

Fruits and veggies provide vitamins and minerals naturally, and in most cases, the body better absorbs vitamins and minerals contained in foods than those contained in a bottle. What's so important about vitamins and minerals? Minerals are inorganic elements — the salt of the earth so to speak. Minerals provide the atoms that form molecules your body needs for metabolism. Vitamins are crucial for proper metabolism, which fuels your body. Your body can't produce vitamins, so it needs to derive vitamins from foods. So why not go ahead and eat foods that contain vitamins instead of eating nutritionally poor foods and then supplementing your diet with vitamins? It makes sense. Plus, for the nutrition they provide, fruits and veggies are low in calories and fat.

The fiber from fruits and vegetables help to keep your digestive system functioning at its best. Fiber gives the digestive system a workout; it keeps the muscles in the walls of the digestive organs strong, and the "roughage" from the fiber helps clear the digestive tract of waste. If waste lingers in the digestive system, it can damage cells. Damaged cells can undergo bad changes that make them cancerous. Eating enough fiber every day can help prevent the development of cancer in places like your colon. Also, fiber fills you up so that you don't have the urge to indulge in more nutritionally poor foods. Eating fiber helps control your appetite: The longer that you remain satisfied between meals, the less likely that you are to snack.

In addition to tasting good, fruits and veggies come in plenty of pretty colors — what other reason do you need to choose them over junk food?

# Exercise Regularly (Not Sporadically)

Many of us are guilty of starting an exercise program (or joining a gym) with good intentions of getting fit and trim. After a while, the novelty wears off, and you let other activities take precedence over exercise. With how busy everyone is these days, it's no surprise that exercise gets bumped off the "To Do" list. Exercise really has to be a priority (like taking a shower every day) for you to incorporate it into your daily routine.

Perhaps your exercise program turned out this way: You're playing a sport or working out at the gym, and you get hurt a bit. You rest at home for a while, and rightly so. The problem is in deciding that hanging out at home is pretty nice. Sure the couch is more comfortable than that stationary bike seat. But, after a while, you start to feel a little sluggish, tired, out of shape, and you drag yourself outside for a walk. That feels better, so you head back to the gym to make another attempt at getting in shape. Then, after a few weeks of working out at full intensity, you hurt yourself again, and the cycle continues.

Break the cycle! Start out slow so you don't injure yourself right off the bat, and then keep it up. Make it part of your wake-up routine. Get up out of bed, go to the bathroom, brush your teeth, drink a glass of water, and exercise. Now, you can take your shower. Studies show that morning exercisers tend to be more habitual exercisers who stick with it. Making a habit out of exercise is what you need to do. (If you're going to have a habit, and all of us do, make it a good one.)

Exercise increases your heart rate, which gets blood flowing faster. The faster that blood flows, the faster you get rid of waste like carbon dioxide; therefore, you have to breathe faster to exhale the carbon dioxide. Breathing faster means you take in oxygen faster, and that speeds up your metabolic reactions. This chain of events is why aerobic (using oxygen) exercise burns glucose and fat. By removing excess glucose and fat from the body, you're helping to reduce the negative effects of diabetes and heart disease. By dilating your arteries and getting your heart pumping, you're strengthening the muscle tissue in your cardiovascular system and preventing plaque from building up and blocking your blood vessels. Exercise builds muscle and bone tissue, so it can reduce or prevent the loss of these tissues later in life. Therefore, your muscles and bones can maintain their ability to move freely — "use it or lose it" is the truth!

# Slather on the Sunscreen

Using a broad-spectrum sunscreen is one of the best ways to prevent skin damage from the sun, including skin cancer. Broad-spectrum sunscreens keep out damaging ultraviolet A (UV-A) rays, which cause wrinkling and sagging, and harmful UV-B rays, which can cause cancer. The best way to avoid skin damage would be to stay out of the sun altogether, but sunlight is beneficial for a few things. Sunlight helps to make vitamin D in the body, which is necessary for strong bones and the balance of calcium and phosphorus. Sunlight also helps to minimize depression in those people who experience seasonal affective disorder (SAD). But, other than the few minutes per day needed to achieve those benefits, the rays of the sun can be harmful. A "healthy looking tan" doesn't exist. Tans are evidence of damaged skin, whether they come from the sun itself or a machine. Wear sunscreen year-round, not just during the summer, and not just when you're going to be at the beach or pool, or playing or watching sports. You are exposed to the sun's rays while driving or just sitting by a window. Get a big bottle of a good, broad-spectrum sunscreen and use it a couple of times every day when you're outdoors.

# Get Seven to Nine Hours of Sleep

Many, many people say they wish that they could get more sleep. Sleep isn't elusive for most people, it's just postponed by TV watching and overbooking each day. If you could skip the late news and the late show or the even-later show, you could easily fit in the right amount of sleep for you. How do you know what the right amount is? By doing it! Leave the television off for a night — I know you can do it, and I bet you'll even enjoy it! — and go to bed as soon as you start to feel tired. When you start yawning, head for bed, and turn your alarm clock off! I know this sounds scary, but trust me, you'll wake up, even without an annoying buzz, ringing, or a DJ. Let yourself fall asleep, and sleep until you wake up all on your own.

Make a note — mental or otherwise — as to how long you slept. If you are truly sleep-deprived, you may sleep longer than you really need for a few nights just because you are starting from behind. But do this exercise for several nights or weeks in a row, and eventually you'll start falling asleep at a certain time and waking up at a certain time. The number of hours between those two times is the right amount for you. You may even find that you never need to set an alarm clock, and that just by waking up on your own without an electronic gadget that you start your day off better. You'll probably feel more rested, refreshed, and ready to handle the stresses of your day. And you'll be doing your body a favor.

The body needs sleep in order to heal itself. Cell repair and growth is much more active at night while you're resting. When you are active, your cells are busy metabolizing to provide energy to keep you going. But they need a break to maintain their own equipment! Sleep allows your brain to recharge. (Although the reasons for why sleep is necessary are unknown.) Sleep also allows your cells to repair and you to relax. Without adequate sleep, people become fatigued and sick. When deprived of sleep for too long, people can die. Sleep is necessary for health, and you can get what you need if you don't fill up every night with an activity that runs until your bedtime or spend hours glued to the TV.

# Relax

If you don't get enough sleep (see the "Get Seven to Nine Hours of Sleep" section in this chapter), maybe it's because you don't allow yourself to relax. Personally, I am guilty of this. I always felt that if I sat down and relaxed, I was wasting time or being lazy. But, as I've grown older, I've discovered that if I *don't* sit down and relax, I am wasting my energy. When I don't take some time to relax, I get so stressed out that I move in perpetual motion but don't get as much accomplished. When I do take the time to relax — even if it's just sitting in a quiet room for ten minutes with my eyes closed — I recharge enough to actually be effective as I expend energy. There's nothing wrong with taking a nap, and a nap doesn't have to be two to three hours long.

I have found that just letting my muscles relax and letting the tension melt away for a short time does wonders for my health. I am less "wired" at night, so I fall asleep easier, I am more patient with my children, nicer to my husband, and my back muscles don't spasm (like they do when I'm stressed out).

Relaxation also reduces blood pressure, which is extremely beneficial if you have hypertension. Not only do the skeletal muscles relax, the smooth muscles of the blood vessels and digestive organs also relax, which can help reduce the risk of several health problems (such as hypertension, diarrhea, and stroke). People relax in different ways; you have to find what works for you. Listening to classical music, reading a book, having a cup of tea (decaf, of course), walking, lying down, yoga, holding a child, petting an animal, taking a bath, and even having sex, are all great ways to relax. Try relaxing and enjoying the simple things in life.

# Eat Oatmeal and Other Grains

Consuming oatmeal can reduce cholesterol in the bloodstream, which helps reduce your risk of heart disease. This reason alone is worth having oatmeal for breakfast every day. But whole grains are also excellent sources of fiber, B-complex vitamins (which help your body use energy from foods better), vitamin E (prevents cell damage from oxidation), and minerals. Whole grains simply are good for you. Oatmeal and grains fill and satisfy you, helping to curb your appetite, making them easy to enjoy when you are counting calories.

# Wash Your Hands

Washing your hands should already be a habit of yours. Hand washing is the best way to prevent the spread of germs — the little buggers that make you sick. You can easily pick up bacteria and viruses if you bite your nails, rub your eyes, pick your nose, or scratch your ear. All those bad habits can put bacteria and viruses right into your system through the tissues and openings in those body parts. Dirty hands can pass the germs to other people or to objects that other people will soon touch.

You can use antibiotics to fight bacterial infections, but they're becoming overused and less effective in some cases. Do your part to shore up the antibiotic choices: Help prevent others from getting infections by washing your hands. Doing so is not only polite and sanitary, it's the right thing to do for your health and the health of others. When you wash your hands, do it with plenty of warm water and soap. You also need to do it long enough to be effective, at least 20 to 30 seconds — just long enough to hum "Happy Birthday" to yourself. Careful now, you may actually find yourself singing in a public restroom.

And speaking of public restrooms, take care not to touch too many surfaces in them. To avoid picking up more germs than you wash off, use a paper towel when turning the water faucet on and off. And if you can't push the door open with your shoulder, hip, or foot when leaving, then turn the door-knob while holding a paper towel, too. So many people don't wash their hands after going to the bathroom that the knob of a restroom door can be teeming with microscopic "wildlife."

Of course, if you are somewhere without water (such as camping or changing a messy diaper along the side of a highway), plenty of antibacterial products (wipes, gels, and lotions) are available that you can keep on hand and use in a pinch. But on a daily basis, consumer tests have shown that regular old soap and water do just as good a job at killing bacteria as do the more expensive antibacterial products. Plus, some scientists are concerned that overuse of antibacterial products may lead to resistant bacteria.

# Do Self-Exams on Breasts or Testicles

Both men and women need to examine themselves monthly for the presence of any lumps or changes in the tissue of breasts or testicles. Testicles move freely within the scrotum, so it is easy to palpate them for any hard tissue or lumps. Although testicular cancer is much more rare than breast cancer, don't let that deter you from performing regular self-examinations. Being familiar with your own body helps you detect any changes more easily.

Breast exams are best performed one week after the menstrual period, and it also helps if the woman is in the shower when doing it. The water helps the fingers to be more sensitive and glide over the tissue easily. A woman can also perform a breast self-examination while lying down. Many women experience fibrocystic changes in which the breast can feel lumpy, full, or tender. Generally, fibroid cysts are fluid-filled and move more freely in the breast, whereas benign or malignant tumors are more solid and less mobile. To be safe, however, go to your doctor if you detect a lump or any changes in the breast tissue. Women should have regular mammograms starting between the ages of 35 and 40.

# Get Regular Check Ups

You and your physician are a team, and both of you should be trying to keep your body at its optimal health. It's so much easier (and more cost-effective) to prevent illness than to diagnose and treat illnesses. But, if you have an illness, don't despair. With proper medical care and genuine attempts to make yourself well (including taking medication as prescribed and following many of the tips in this list), most illnesses can be overcome. The body has an amazing capacity to heal itself, and with your healthcare provider's help, you should be able to restore, improve, or maintain good health. Keep track of your medical visits, test results, immunizations, medications you were prescribed, and suggestions your physician made during your appointments. That way, you and your physician can look back through your records periodically and see what patterns may be developing; you can adjust your "game plan" to make your best offensive and defensive moves against disease.

# Chapter 17

# Ten Great Anatomy & Physiology Web Sites

● ● ● ● ● ● ● ● ● ● ● ● ● ● ● ● ● ● ● ● ● ● ● ● ● ● ● ● ● ● ● ● ● ● ● ● ● ● ●

*In This Chapter*

▶ Surfing for super anatomy and physiology info

▶ Looking for self-quizzes, animations, and close-up looks at the body

● ● ● ● ● ● ● ● ● ● ● ● ● ● ● ● ● ● ● ● ● ● ● ● ● ● ● ● ● ● ● ● ● ● ● ● ● ● ●

*I*f you have an interest in biology, and the body simply amazes you, check out these sites that I found on the worldwide Web. Of course, these aren't the only ten anatomy-related Web sites, but they're ones that I thought may interest or help somebody who's just learning about how the body works. Hope these sites give you a "leg up" on anatomy!

## Anatomy and Physiology

www.msms.doe.k12.ms.us/biology/anatomy/apmain.html

Anatomy and Physiology is a site intended for high school students who are just learning these subjects. The reviews cover the body, part by part, and each system of the body is represented. When you click on one of the photos, you get a review of each system and the major structures. Navigation is easy, with a menu running along the side of each page. You also get sound effects (listen to the digestive system at work!) and color illustrations. The site is 5 years old, but the human body hasn't changed at all since then. The information and photos are still good.

## Human Anatomy and Physiology Case Study Project

http://faculty.niagara.edu/bcliff/

The Human Anatomy and Physiology Case Study Project is a site run by the biology faculty at Niagara University. The case studies help anatomy and physiology students integrate the knowledge that they gain in their course with real-life applications. The case studies develop problem-solving, analytical, and diagnostic thinking skills. The studies are systematic, just like going through the diagnosis and treatment for a patient. This site is fun and challenging, if you're going the pre-med route.

# Merck.com

www.merck.com/disease/heart/

Pharmaceutical and research giant Merck Co. offers an excellent resource for information on your heart and how to keep it healthy. You'll find pictures and information on coronary health, risk factors, and lifestyle. Take the "Heart Smart Trivia" quiz or check out the adult guide to proper eating. Also, this site provides you with a handy record-keeping section with printable forms and schedules for things like insulin injections, medical information, and blood-pressure readings. Just ignore the annoying pop-up ads.

# Linkpublishing.com

www.linkpublishing.com

Texas-based Link Publishing has an excellent educational Web site. In addition to ordering their products, and seeing samples from their textbooks and laboratory workbooks, you can take an online examination for every topic in anatomy and physiology. These online exams, made up of photographic identifications, matching questions, and multiple-choice questions are great ways to review. Test your knowledge of concepts, structures, and functions. After you submit your answers, you immediately receive your "grade" and an explanation of answers. The feedback shows you which of your answers are correct and which ones are wrong. Best of all, this service is free!

# Tangentscientific.com

www.tangentscientific.com

If you want to shell out a few bucks for some online e-learning materials or software to help you study anatomy (and other science subjects), check out the material available through Tangent Scientific Supply. The company's science education software includes several choices for anatomy and physiology, including A.D.A.M. — the virtual dissection. A.D.A.M. shows the human body layer by layer. More than 18,000 photographs were taken throughout the dissection of a real human cadaver, and they were then catalogued and made available to anatomy students as A.D.A.M. These photographs are useful if you are going to be studying anatomy and physiology at the advanced level, or if the smell of formaldehyde is getting to you and you need a break from your own cadaver. The e-learning courses from Tangent enable you to really get into anatomy and physiology for an entire year. For about $200, you can get a 12-month subscription to one of their several online anatomy products featuring A.D.A.M.

# Bio201

www.gwc.maricopa.edu/class/bio201

The Bio201 Anatomy and Physiology Web page features the anatomy and physiology course at GateWay Community College in Phoenix, Arizona. The site provides outlines, study questions, "survival tips," essay help, a "toolbox" for success in a science course, and interactive tutorials (in JavaScript) — check out CyberHeart in the cardiophysiology tutorial. But, my favorite part of this site is the "Cool Biomedical Site of the Moment," which provides a link to a different health-related site every time you click on it. Some of the links include "JAMA News," from the *Journal of the American Medical Association*; "Healthy News Daily," which provides health news that's updated daily and information on alternative medicine; "CNN-Health," which is the health channel of CNN.com; "Mayo Medical Daily News," from the Mayo Health Clinic, and "Heart Healthy." You can register to receive daily updates; check out the listing of Web search sites, as well. Way cool.

# University of Arizona Physiology Department

http://server.physiol.arizona.edu/Physiology/Instruct/
LectureNoteArchive.html

If you need some good old-fashioned lecture notes, check out the extensive listing from the University of Arizona Physiology Department. In addition to current lecture notes in each subject area of physiology, you'll find an archive that provides lecture notes back as far as 1994. In addition to lecture notes, check out the practice exams, study guides, problem sets, photos of real body parts, animation, and 3D structures. Pull up a chair and pretend you're in class!

# About.com

`http://biology.about.com`

Need some homework help? A learning guide? Science fair project? This is the place to go! At About.com's Biology page, you can see virtual dissection from cats to rats, simulations, tutorials on different systems and structures of the body, and links to the *Atlas of Human Anatomy*. Take a look at magnetic resonance imaging (MRI) and computed tomograph (CT) scans and the "digital anatomist." I like the "Did You Know" archive of interesting facts about the human body and the listing of biology prefixes and suffixes, so you can figure out complex terms all by yourself, like: *pneumonoultramicroscopicsilicovol-canoconiosis,* which is a real word, by the way.

# Innerbody.com

`www.innerbody.com`

Innerbody.com runs "Human Anatomy Online," which provides pictures of ten systems of the body. When you click on the pictures, you'll get a bigger picture covered with little green and red markers. You can quiz yourself by guessing what is under the green or red marker before you move your mouse over it and reveal the label. The site offers animations, images, and descriptions for nearly every structure or system in the body.

# Madsci.org

`www.madsci.org/~lynn/VH/annotated.HTML`

If you need to see part of the human body, animated or still, check out the Visible Human Project, by Lynn Bry, MD, PhD, of Brigham and Women's Hospital (an affiliate of Harvard Medical School, Boston, Massachusetts). She used the collection of 18,000+ digitized photos from a human cadaver at the National Library of Medicine (NLM) and Washington University in St. Louis, Missouri, to provide an animated and annotated tour through the human body. Although the NLM photos are a great collection, they don't necessarily tell you what you're looking at. Dr. Bry has animated and annotated many of these photos, providing all of us with a valuable learning tool. Thanks, Dr. Bry! And if you have a few minutes, download the "flythrough" of the colon of the Visible Human — it's quite a trip.

# Index

• *C* •

**• E •**

### • F •

# FOR DUMMIES®

## The easy way to get more done and have more fun

---

## PERSONAL FINANCE & BUSINESS

**Investing FOR DUMMIES**
0-7645-2431-3

**Home Buying FOR DUMMIES**
0-7645-5331-3

**Grant Writing FOR DUMMIES**
0-7645-5307-0

**Also available:**

Accounting For Dummies
(0-7645-5314-3)

Business Plans Kit For Dummies
(0-7645-5365-8)

Managing For Dummies
(1-5688-4858-7)

Mutual Funds For Dummies
(0-7645-5329-1)

QuickBooks All-in-One Desk Reference For Dummies
(0-7645-1963-8)

Resumes For Dummies
(0-7645-5471-9)

Small Business Kit For Dummies
(0-7645-5093-4)

Starting an eBay Business For Dummies
(0-7645-1547-0)

Taxes For Dummies 2003
(0-7645-5475-1)

---

## HOME, GARDEN, FOOD & WINE

**Feng Shui FOR DUMMIES**
0-7645-5295-3

**Gardening FOR DUMMIES**
0-7645-5130-2

**Cooking FOR DUMMIES**
0-7645-5250-3

**Also available:**

Bartending For Dummies
(0-7645-5051-9)

Christmas Cooking For Dummies
(0-7645-5407-7)

Cookies For Dummies
(0-7645-5390-9)

Diabetes Cookbook For Dummies
(0-7645-5230-9)

Grilling For Dummies
(0-7645-5076-4)

Home Maintenance For Dummies
(0-7645-5215-5)

Slow Cookers For Dummies
(0-7645-5240-6)

Wine For Dummies
(0-7645-5114-0)

---

## FITNESS, SPORTS, HOBBIES & PETS

**Fitness FOR DUMMIES**
0-7645-5167-1

**Golf FOR DUMMIES**
0-7645-5146-9

**Guitar FOR DUMMIES**
0-7645-5106-X

**Also available:**

Cats For Dummies
(0-7645-5275-9)

Chess For Dummies
(0-7645-5003-9)

Dog Training For Dummies
(0-7645-5286-4)

Labrador Retrievers For Dummies
(0-7645-5281-3)

Martial Arts For Dummies
(0-7645-5358-5)

Piano For Dummies
(0-7645-5105-1)

Pilates For Dummies
(0-7645-5397-6)

Power Yoga For Dummies
(0-7645-5342-9)

Puppies For Dummies
(0-7645-5255-4)

Quilting For Dummies
(0-7645-5118-3)

Rock Guitar For Dummies
(0-7645-5356-9)

Weight Training For Dummies
(0-7645-5168-X)

---

**Available wherever books are sold.**
**Go to www.dummies.com or call 1-877-762-2974 to order direct**

**WILEY**

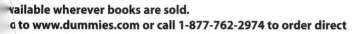

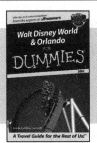

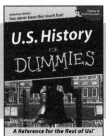
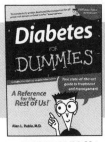

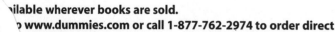